Superconductor Engineering

Elmer L. Gaden, Jr., series editor
Thomas O. Mensah, volume editor

B.F. Allen
U. Balachandran
Osman A. Basaran
A. Bhargava
J. Brynestad
Charles H. Byers
J.D. Connolly, Jr.
Lawrence P. Cook
S.E. Dorris
J.T. Dusek
N.M. Faulk
J.P. Formica
K. Forster
Ahmed M. Gadalla
N.A. Gokcen
K.C. Gorretta
A.S. Gurav
Michael T. Harris
Turi Hegg
H. Hsu
J.R. Hull
Sendjaja Kao
T.T. Kodas
Paisan Kongkachuichay

Patricia S. Kraft
D.M. Kroeger
M.T. Langan
Sangho Lee
N.F. Levoy
S.-C. Lin
D. Luss
K.Y. Simon Ng
Warren H. Philipp
Julia M. Phillips
J.J. Picciolo
R.B. Poeppel
J.T. Richardson
M.A. Rodriguez
T.E. Schlesinger
J.T. Schwartz
Timothy C. Scott
R. Semiat
J.P. Singh
S. Sundaresan
R.L. Snyder
Lawrence Suchow
T.L. Ward
Winnie Wong-Ng

C.A. Youngdahl

AIChe Staff
Maura N. Mullen, Managing Editor
Arthur H. Baulch, Editorial Assistant
Cover Design: Mark J. Montgomery

**Inquiries regarding the publication of Symposium Series Volumes should be directed to
Dr. Elmer L. Gaden, Jr., series editor,
University of Virginia, Department of Chemical Engineering,
Thornton Hall, Charlottesville, Virginia 22903-2442. FAX (804) 924-6270.**

AIChE Symposium Series

Number 287   1992   Volume 88

Published by

American Institute of Chemical Engineers

345 East 47 Street   New York, New York 10017

Copyright 1992

American Institute of Chemical Engineers
345 East 47 Street, New York, N.Y. 10017

AIChE shall not be responsible for statements or opinions advanced in papers or printed in its publications.

Library of Congress Cataloging-in-Publication Data

Superconductor engineering / Thomas O. Mensah, editor.

     p. cm. — (AICHE Symposium series; no. 287, v. 88 (1992))

    Includes index.

    ISBN 0-8169-0567-3

    1. Superconductors—Chemistry, 2. High temperature superconductors. 3. Chemical reaction, Rate of. I. Mensah, Thomas O., 1950–   . II. Series.
QC611.97.C54S86   1992
621.3—dc20   92-9161
                                                            CIP

FOREWORD

This book contains papers presented at National AIChE meetings under the auspices of the Materials and Engineering Sciences Division. This effort has the goal of keeping Chemical Engineers at the forefront of advanced materials processing as well as new technology.

Contributors from several leading laboratories like NASA, Argonne, NIST, AT&T Bell Laboratories, Oak Ridge, Sandia, and key universities allow the reader to see various approaches to solving the main challenge facing the industry, namely, how to make products that maintain strength and quality as well as performance. Industrial participation by companies like DuPont, AT&T and other consortia in place around the world can accelerate development as can be seen in the text.

In this book the requisite chemical engineering effort is divided into three sections. The first section deals with the fundamental chemistry, superconductor thermodynamics and reaction kinetics. Synthesis and chemistry of high Tc superconductors are treated mostly in this section. The next section focuses on the challenges of processing, fabrication and engineering scale-up. Several fabrication methods are examined. These include bulk powder processing as well as thin film fabrication. Transformation to wire by extrusion and aerosol process for making high Tc materials are also discussed.

Finally the third section deals with characterizing the product using sophisticated techniques like XRD. Properties of high Tc superconductors measured *in situ* can aid process parameters development for automation and process control, key elements for achieving product quality. The challenge of producing large quantities of high Tc materials that preserve mechanical strength and integrity without sacrificing performance of high critical currents in large magnetic fields must be overcome. Transport phenomena analysis and advanced process scale-up analysis will be required as we continue to push the technology toward commercialization. Chemical engineers can play a significant role as demonstrated in this book.

In attempting to bring such frontier subjects to the mainstream of chemical engineering I have been fortunate to receive support from several of my colleagues, Dr. Gideon Grader, AT&T Bell Labs; Professor Dan Luss, University of Houston; Professor Jimmy Wei, Dean of Engineering, Princeton University; and members of the executive committee of MESD, as well as other leaders in AIChE. I thank them all for their help. I am grateful to all the contributors from various research laboratories and universities in the U.S. for their enthusiastic support.

Thomas O. Mensah, Ph.D.
Volume Editor
Supercond Technology Inc.
Aerospace Materials Div.
Mitchell Boulevard N.E.
Norcross, GA 30091-2883

CONTENTS

Introduction

Historical Perspective

As we move into the 21st century, the next wave of technological innovation in advanced materials processing will involve high temperature superconductors. In fact, the discovery of a new inorganic superconductor by IBM scientists led to one of the fastest awards of Nobel prizes in physics, that of Bednorz and Mueller in 1986. However the phenomenon of superconductivity has been known since 1911, when Kammerlingh Onnes discovered that mercury, when cooled below its critical temperature 4.2 degrees kelvin, exhibited zero resistance to electricity. At the same time it causes the metal to expel all magnetic flux from its interior. The latter property is called the Meisner effect.

However, for the next several decades, until the IBM announcement, the highest critical temperature was around 23 K observed in various metals and alloys. These have been referred to as Type I superconductors. Since then a flurry of activity in various laboratories around the world has pushed the critical temperature to well above liquid nitrogen temperature of 77 K. Critical temperatures of 127 K, and above, have been cited in materials (mostly oxides) commonly known as Type II superconductors. Such materials are called high Tc materials to differentiate them from Type I materials which are mostly metals and alloys as mentioned above.

The theory used to explain Type I superconductivity is the BCS (Bardeen Cooper Shrieffer). New theories are being proposed for the new high Tc materials. But at the moment experimentalists are further ahead than the theorists in predicting properties of these materials.

Application

The application of superconductivity in modern society can revolutionize everything from high speed computing to magnetically levitated high speed trains. In ships and submarines the use of superconducting magnets employing magnetohydrodynamic forces may provide quieter underwater vehicles that are difficult to detect.

Applications in medicine include, Computer Aided Tomography (for imaging the human brain) to SQUIDS (Superconducting Quantum Interference Devices) that can even measure brain waves (such as, ultra low magnetic fields). The need to use cryogenic fluids like Liquid Helium (4.2 K) in order to reach the critical temperatures in alloys like niobium titanium present serious design challenges for engineers. In fact, the largest scientific instrument under development, the SSC or the Superconducting Super Collider will use this approach, where Rutherford-type superconducting cables of niobium titanium alloy in a copper matrix are utilized to wind the magnets for the 53 mile long particle accelerator. Two beams of protons will be accelerated at nearly the speed of light in opposite directions in this 53 mile ring so that at the moment of collision they will release enormous energy, approximately 40 trillion electron volts. This energy will exceed the instantaneous output of all power plants on earth compressed in the space smaller than a single proton. New elementary particles beyond quarks and leptons could be created and detected using sophisticated detectors and computers. The Superconducting Super Collider could be the largest project in American science in this decade. The project is moving from design and conception phase to engineering construction. The price tag on the SSC ($8.5 billion) can be reduced if superconductors that operate at room temperature are available. For one thing cryogenic cooling and the engineering cost of operation at 4.2 K will be reduced, not counting the possibility to design a particle accelerator of smaller size. The importance of the new superconductors Type II which will exhibit zero resistance at higher temperatures is appreciated when one thinks of common application like electric power generation, storage and transmission which could reduce the average utility bill.

Engineering Challenge

High Tc superconducting materials must be produced at a cost competive level while maintaining the qualities of strength and high current carrying densities under imposed magnetic fields of 6 Tesla. This is where chemical engineers have a unique but special role in developing appropriate cost effective processes using the new design pradigm of zero pollution.

Thomas O. Mensah, *volume editor*

KINETICS OF YBa$_2$Cu$_3$O$_6$ FORMATION VIA *IN SITU* XRD

J. P. Formica, K. Forster, J. T. Richardson, and D. Luss ■ Department of Chemical Engineering, University of Houston, Houston, TX 77204-4792

The reactions leading to the formation of YBa$_2$Cu$_3$O$_6$ have been followed *in situ* using high-temperature, time-resolved X-ray diffraction. The time dependence of the mass fraction of this material and other phases was quantitatively determined in nitrogen, helium, and air at 750 °C. In nitrogen and helium, BaCu$_2$O$_2$ forms rapidly initially, and its rate of reaction with Y$_2$O$_3$ controls the time needed to complete the formation of YBa$_2$Cu$_3$O$_6$. In air, additional phases are formed, reducing the rate of production and amount of YBa$_2$Cu$_3$O$_{7-x}$.

Since the discovery of the new class of ceramic superconductors [1-4], there has been much interest in mass production of these materials. In the case of the YBa$_2$Cu$_3$O$_{7-x}$ (123) system, this interest has increased with the advent of high J$_c$ thin films and melt-textured materials which are required for device development. This has resulted in the application of many different ceramic processing techniques to achieve high production rates for high-grade 123 ceramics. In the optimization and scale-up of these processes, it is essential to know the reaction pathways, decomposition kinetics of the precursor materials and subsequent reactions to form the final product.

Traditional techniques for the study of solid state reactions include thermogravimetry (TG), differential thermal analysis (DTA), and powder X-ray diffraction (XRD). A TG and DTA study by Jiang et al. [5] suggested that tetragonal 123 is formed in air between 800 °C and 950 °C by the two-step process:

$$BaCO_3 + CuO \rightarrow BaCuO_2 + CO_2 \qquad (1)$$

$$Y_2O_3 + 4BaCuO_2 + 2CuO \rightarrow 2YBa_2Cu_3O_6 + \tfrac{1}{2}O_2 . \qquad (2)$$

The formation of the YBa$_2$Cu$_3$O$_6$ is reported to be controlled by a second order chemical reaction.

Gadalla and Hegg [6] performed a similar TG and DTA study, and in addition to Reaction (1), suggested the following formation reactions:

$$Y_2O_3 + BaCuO_2 \rightarrow Y_2BaCuO_5 \qquad (3)$$

$$Y_2BaCuO_5 + 3BaCuO_2 + 2CuO \rightarrow 2YBa_2Cu_3O_{7-x} . \qquad (4)$$

Using classical kinetic models to interpret the data, they concluded that these overlapping reactions are diffusion-controlled. Wu et al. [7] questioned the validity of Gadalla and Hegg's analysis since it was applied to overlapping reactions. Wu et al. also examined 123 formation in air from BaCuO$_2$, Y$_2$BaCuO$_5$ (211) and CuO. They reported good agreement between kinetic parameters calculated from TG and DTA data and those obtained from

X-ray data. A shrinking core model was proposed to explain the particle-size dependence of the formation reaction. However, to calculate concentrations from X-ray data, Wu et al. assumed that the majority of the mass was in the 123 and 211 phases, and did not account for other phases, which may be correct when the conversion levels are high but not in the initial stages of the reactions.

Ruckenstein et al. [8] prepared 123 (ostensibly in air) from combinations of oxides and $BaCO_3$. Starting materials were ground together, fired in alumina boats at 940 °C for different lengths of time, quenched to ambient temperature, then X-rayed. They concluded that $BaCO_3$ decomposition is the rate-limiting reaction and that the 123 phase can be rapidly obtained when $BaCuO_2$ is used as the barium source. The height of the strongest diffraction peak was taken as a measure of phase concentration. They claimed a general proportionality between peak area and height, then used the peak height alone as a measure of concentration. The peak width is directly related to the crystallite size, and during these reactions, all crystallites change size. In the case of small crystallites, this size effect may lead to a large error if the concentration is determined solely from the peak heights.

High-temperature X-ray studies complement the traditional analytical techniques by allowing the observation of metastable high-temperature phases. Furthermore, TG and DTA results require knowledge of the phases present at a given time or temperature for their interpretation. High-temperature X-ray studies can provide this lacking structural information. Coupled with a fast detector, a high-temperature diffraction system can examine the evolution of crystalline phases with time.

There are a number of reported high-temperature X-ray studies of the formation and decomposition reactions in the 123 system. Dubrovina et al. [9] examined the behavior of the 211 phase, Y_2BaCuO_5, in air. The formation and reactions of intermediary binary cuprates were studied by Chigareva et al. [10]. Kulpa et al. [11] studied the reduction of 123 in hydrogen. The formation of 123 in air was studied by Lin et al. [12] who reported the appearance of the tetragonal phase at 880 °C. Thomson et al. [13] studied phases formed in helium and air from precursors prepared to minimize macroscopic diffusion limitations. The tetragonal phase was reported to form in helium at temperatures as low as 607 °C. Thomson et al. used relative intensities (ratios of peak heights) as a guide to concentrations.

The present work describes a high-temperature X-ray diffraction study of the kinetics of formation of tetragonal $YBa_2Cu_3O_6$. Measurements were made in air, nitrogen, and helium at 750 °C starting from a commercial, spray-roasted precursor (produced by SSC, Inc.) containing Y_2O_3, $BaCO_3$ and CuO. The mass fractions of the reaction products were determined by a Rietveld refinement of the diffraction patterns of quenched samples. This method uses structural information for each of the phases to compute their diffraction patterns and performs a least-squares minimization of the difference between the measured and calculated patterns.

EXPERIMENTAL SYSTEM AND PROCEDURE

In situ high-temperature, time-resolved X-ray diffraction experiments were performed with a Siemens D5000 diffractometer in the vertical $\theta/2\theta$ configuration. Copper $K\alpha_1$ radiation was generated using a 1.5 kW, sealed-tube source and a curved germanium (111) incident beam focusing monochromator. A Bühler HDK 2.3 high-temperature stage ("hot stage") with an environmental heater held the samples at elevated temperatures under controlled atmospheres. A Dycor MA200M quadrupole gas analyzer was used to monitor the composition of the gas phase in the hot stage. Gas was sampled via a fused silica capillary positioned approximately 10 mm above the powder sample. The diffracted intensities were collected with a Braun OED-50M position sensitive detector (PSD) which can receive data simultaneously over a 5.26° range in 2θ. High-precision data can be collected with the PSD roughly one hundred times faster than with a standard scintillation detector, enabling time-resolved studies.

Powder samples were mounted on a gold-plated platinum heating strip approximately 50 mm long, 12 mm wide and 0.15 mm thick. The illuminated area of a sample was at most 4 mm by 12 mm. Gold-plating prevented the formation of barium platinum oxides at elevated temperatures. A type S (Pt/Pt-10Rh) thermocouple was spotwelded to the underside of the heating strip.

The precursor was suspended in absolute ethanol and sprayed with an artist's airbrush (Paasche VL) onto the heating strip. The ethanol evaporated under an infrared lamp leaving a thin solid coating. The thermal conductivity of the film is generally low, so the temperature of the diffracting material near the sample surface differs from the controlled ribbon temperature. To minimize the temperature gradient, the sample strip was surrounded by a nearly cylindrical platinum/rhodium alloy heater roughly 30 mm in diameter from which the upper ~30° arc was removed to allow the unrestricted passage of the X-ray beam. While a thick coating maximizes the diffracted intensity, it also creates a large thermal gradient across the film. A compromise thickness was determined, and the data was corrected for this reduced sample thickness as described in Appendix A. A room-temperature scan was then run while purging the hot stage with the desired gases. The gas analyzer was used to determine that two hours are adequate for this purging process.

The precursor material prepared by Seattle Specialty Ceramics (SSC), Inc. consisted of an intimate mixture of $BaCO_3$, CuO, Y_2O_3, and a trace amount of $Ba(NO_3)_2$. Electron microprobe analysis showed this precursor powder to be homogeneous at the particle level. The homogeneity of the powder is crucial for these experiments, because the heating strip is thinly coated with sample, and interparticle mass transfer is slow.

The high-temperature data was collected in the fixed scan mode, i.e. the PSD was stationary. The most intense reflections from the 123 phase are the (103) and the (110) which occur at ~32.4° and ~32.8° in 2θ, respectively. By examining these strong reflections, data can be collected rapidly with good counting statistics. Most materials expand when heated, shifting the diffraction peaks to lower angles; in the case of the 123 material, the shift is roughly 0.5° in 2θ at 750 °C. Other phases of interest here are the barium copper oxides, as well as $BaCO_3$. By setting the detector to examine the 2θ range from 28.5° to 33.5°, strong reflections from most phases of interest are observed.

The data was stored in a Braun MCA 3/1 multichannel analyzer (MCA). Each channel in the MCA corresponds to a 2θ angular range of 0.05°. After filling all 8192 channels or when the data collection finished, the MCA transferred the collected intensities to a host MicroVAX II computer. A typical high-temperature experiment began with a 60 sec room-temperature fixed scan over the angular region of interest. The sample was then heated at 20 °C/sec to the desired temperature. Forty successive 60 sec fixed scans were then taken with a 2 sec delay between each scan. Another thirty scans were made with 10-30 min delays between each to allow the reactions sufficient time to run to completion. The sample was then quenched at a rate of 20 °C/sec. By that time, the sample was primarily

$YBa_2Cu_3O_{7-x}$ ($x \rightarrow 1$ for nitrogen, and helium environments), and it was assumed that little change in the sample would occur during the quenching process. A final, very long room-temperature scan was made over a $100°$ range in 2θ. The final composition of the sample was quantitatively determined using a Rietveld refinement technique on this long scan.

DATA ANALYSIS

The high-temperature data was taken over short time intervals on thin samples, so the counting statistics were somewhat reduced. The peaks of each pattern were fit with a constant background and Gaussian profiles, using a weighted least-squares algorithm. The area under a diffraction peak (integrated intensity) of a given phase is proportional to the number of unit cells of that phase illuminated by the X-ray beam.

For the experiments conducted in nitrogen and helium, the mass fraction of each phase in the diffraction pattern of the quenched sample was determined by a Rietveld refinement. This method fits every point in the diffraction pattern. It requires knowledge of every phase and its structural parameters and is described in Appendix B. Since the mass fraction of the 123 phase is proportional to the area under its diffraction peaks, the integrated intensity from the last high-temperature scan is proportional to the mass fraction determined from the Rietveld refinement. The mass fraction for any other time t is then determined by multiplying the mass fraction obtained from the Rietveld refinement by the ratio of the integrated intensity for time t to that of the last high-temperature scan.

In the experiments conducted in air (a mixture containing 80% nitrogen and 20% oxygen), some of the phases present in the quenched sample were not identifiable from the JCPDS PDF-2 database [14] of X-ray diffraction patterns. Thus, a Rietveld analysis could not be conducted. The mass fraction of the 123 was calculated using the somewhat less accurate procedure described in Appendix B.

RESULTS AND CONCLUSIONS

The time variation of the high-temperature diffraction patterns for samples kept in helium and air atmospheres at 750 °C are shown in Figs. 1(a) and 1(b), respectively. For clarity, only a few scans are shown here, and the angular range has been reduced to highlight the 123 and $BaCu_2O_2$ peaks. The observed peaks are the (110) and (103) reflections of the $YBa_2Cu_3O_6$ phase and the (103) and (200) reflections of $BaCu_2O_2$ (though $BaCu_2O_2$ is only a minor phase in air). It is important to point out that the intensity scales in these two figures are different. Thus, the maximal peak intensity of $YBa_2Cu_3O_6$ in Fig. 1(a) is about three times larger than that in Fig. 1(b). The rapid production of $BaCu_2O_2$ at the early stages of the reactions in the inert environments is surprising. It indicates a different reaction pathway from that reported previously in the literature.

A peak from an unidentified phase exists at 2θ of ~30.5° in Fig. 1(b). Also, the two peaks associated with the 123 tetragonal phase become broad and overlapped in an oxygen environment because of variation in the lattice parameters as oxygen is incorporated into the structure. Both $BaCuO_2$ and Y_2BaCuO_5 have strong diffraction lines between 28.5° and 33.5° in 2θ, and neither phase is present in the samples processed at 750 °C in the inert atmospheres. Neither sample should have reacted or decomposed upon quenching, and no secondary phases were present in the room-temperature pattern. The absence of these phases may be due to the operative reaction pathway. Most reported studies of these phases used

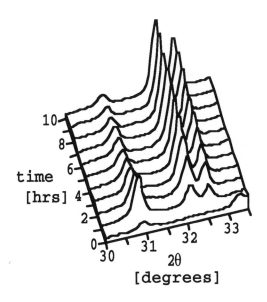

FIG. 1(a): SSC 123 precursor
750°C in helium

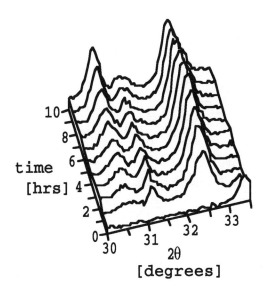

FIG. 1(b): SSC 123 precursor
750°C in air

much slower heating rates than the 20 °C/sec used in this study. It is possible that the formation of these two phases is energetically favored at lower temperatures, but the sample is quickly heated through this regime to a higher temperature where the formation of the $BaCu_2O_2$ is favored.

The mass fractions' time dependence for air, nitrogen, and helium environments are shown in Figs. 2(a) and 2(b). During the first 50 min, the conversion rates in the nitrogen and helium atmospheres are different, and the rate of formation in the helium atmosphere is faster than in nitrogen. However, this difference is short-lived, and after this period, the time dependence of the mass fractions of $YBa_2Cu_3O_6$ in both inert gases is the same within the experimental accuracy.

In air, the conversion rate and final amount of product obtained are rather different from those obtained in the two inert gases. The reaction in air proceeds at a much slower rate than in either nitrogen or helium. From the diffraction pattern taken at the end of the reaction time (10 hrs), we found that in air many phases were

formed that did not form in the nitrogen or helium atmospheres. Some of these phases were unidentifiable even with the aid of the JCPDS PDF-2 database. This required the use of a somewhat less accurate determination of the mass fraction described in Appendix B. The formation of these other stable phases must be the factor that impedes the formation of the $YBa_2Cu_3O_6$.

The high-temperature diffraction data in the inert atmospheres indicate that the following reactions occur:

$$2CuO + BaCO_3 \rightarrow BaCu_2O_2 + CO_2 + \tfrac{1}{2}O_2 \quad (5)$$

$$Y_2O_3 + 4BaCu_2O_2 + \tfrac{3}{2}O_2 \rightarrow 2YBa_2Cu_3O_6 + 2CuO \quad (6)$$

The data suggest that the production of both the 123 phase and $BaCu_2O_2$ is rapid. This indicates that the reactions are initially not diffusion limited. This is probably due to the intimate mixing of the starting material.

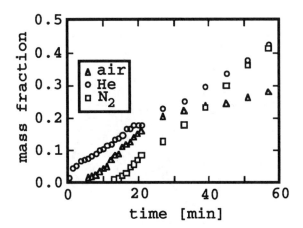

FIG. 2(a): Formation of $YBa_2Cu_3O_6$
at 750 °C (short times)

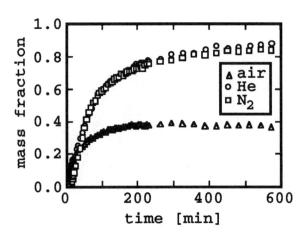

FIG. 2(b): Formation of $YBa_2Cu_3O_6$
at 750 °C (long times)

The main advantage of the high-temperature XRD technique is that it enables a continuous measurement of the concentrations of key phases which appear in the sample during the reaction at a controlled temperature in a controlled environment. This enabled us to determine, for example, the occurrence of Reaction (5), which was undetected by previous investigators and gain an understanding of the cause of the observed difference in the rate of formation of $YBa_2Cu_3O_6$ in air or either nitrogen or helium. We intend to carry out additional experiments at several other temperatures in order to determine the reaction pathways more precisely and to develop a kinetic model. This information is essential for a rational design and scale-up of a calcination process.

The determination of the concentrations of the various reacting phases from the X-ray diffraction patterns is complicated by the presence of phases having rather different structure factors and by the changes in the crystallite size, especially as phases form or disappear. The methods of analysis exploited here are more accurate than those used by previous investigators and should be used in future studies.

A shortfall of the Rietveld method is that phases with extremely small crystallites (<10 nm) and amorphous phases cannot be handled. An apparent deficiency in these phases results. A priori knowledge of the presence of such phases is essential. As an example, in both patterns on which we performed a Rietveld refinement, a mass balance shows that the mass of Y_2O_3 and CuO were underestimated by less than three percent. When the conversion to 123 is low and the contribution of the microcrystalline phases to the total mass is large, this error is expected to increase.

ACKNOWLEDGMENTS

This work is supported by the Texas Center for Superconductivity at the University of Houston under a grant from the Defense Advanced Research Projects Agency and the State of Texas. We thank SSC, Inc. for providing the precursor powder used in these studies. We are also grateful to Robert Von Dreele, Allen Larson, and David Bish of Los Alamos National Laboratory for their assistance with GSAS, and to James Cline of the National Institute of Standards and Technology and William Thomson of Washington State University for their helpful discussions. Special thanks to

Vassiliki Milonopoulou for her help with the profile fitting.

NOTATION

Roman Letters

A_E	= environmental absorption factor
A_V	= volume absorption factor
F	= structure factor
i	= subscript denoting a phase
I	= integrated intensity
LP	= Lorentz-polarization factor
M	= multiplicity factor
N	= number of unit cells
t	= effective sample thickness
V	= unit cell volume
x	= mass fraction
y	= mole fraction
z	= molecules per unit cell

(hkl) = Bragg reflection with Miller indices h, k, and l

Greek Letters

γ	= scaling factor
μ	= linear absorption coefficient
θ	= Bragg angle
θ_m	= monochromator reflection angle

LITERATURE CITED

1. J.G. Bednorz and K.A. Müller, *Z. Phys. B*, **64**, 189 (1986).

2. M.K. Wu, J.R. Ashburn, C.J. Torng, P.H. Hor, R.L. Meng, L. Gao, Z.J. Huang, Y.Z. Wang, and C.W. Chu, *Phys. Rev. Lett.*, **58**, 908 (1987).

3. H. Maeda, Y. Tanaka, M. Fukutomi, and T. Asano, *Jpn. J. Appl. Phys.*, **27**, L209 (1988).

4. Z.Z. Sheng, A.M. Hermann, A. El Ali, C. Almason, J. Estrada, T. Datta, and R.J. Mason, *Phys. Rev. Lett.*, **60**, 937 (1988).

5. Jiang, X.P., J.S. Zhang, J.G. Huang, M. Jiang, G.W. Qiao, Z.Q Hu, and C.X. Shi, *Mater. Lett.*, **7**, 7/8, 250 (1988).

6. Gadalla, A.M. and T. Hegg, *Thermochim. Acta*, **145**, 149 (1989).

7. Wu, N.-L., T.-C. Wei, S.-Y Hou, and S-Y. Wong, *J. Mater. Res.*, **5**, 10, 2056 (1990).

8. Ruckenstein, E., S. Narain, and N-L. Wu, *J. Mater. Res.*, **4**, 2, 267 (1989).

9. Dubrovina, I.N., R.G. Zakharov, E.G. Kostitsyn, A.V. Antonov, V.F. Balakirev, and N.A. Vatolin, *Superconductivity*, **3**, 6, S102 (1990).

10. Chigareva, O.G., G.A. Mikirticheva, V.I. Shitova, S.K. Kuchaeva, L.Yu. Grabovenko and R.G. Grebenshchikov, *Mat. Res. Bull.*, **25**, 1435 (1990).

11. Kulpa, A., A.C.D. Chaklader, D. Roemer, D.L. Williams, and W.N. Hardy, *Supercond. Sci. Technol.*, **3**, 483 (1990).

12. Lin, G.M., Q.Z. Huang, J.X. Zhang, G.G. Siu, and M.J. Stokes, *Solid State Commun.*, **68**, 7, 639 (1988).

13. Thomson, W.J., H. Wang, D.B. Parkman, D.X. Li, M. Strasik, T.S. Luhman, C. Han, and I.A. Aksay, *J. Am. Ceram. Soc.*, **72**, 10, 1977 (1989).

14. *Powder Diffraction File (Level II, Sets 1-40)*, Joint Committee for Powder Diffraction Standards - International Centre for Diffraction Data, Swarthmore, PA (1990).

15. Garn, P.D. and O. Menis, *Certificate: ICTA Certified Reference Materials for Differential Thermal Analysis from 125-940°C*, p. 15, National Bureau of Standards, Washington, DC (1971).

16. Merryman, R.G. and C.P. Kempter, *J. Am. Ceram. Soc.*, **48**, 4, 202 (1965).

17. Larson, A.C. and R.B. Von Dreele, computer code GSAS (Los Alamos National Laboratory, Los Alamos, NM, 1991).

APPENDIX A: TEMPERATURE CALIBRATION

The sample surface temperature was calibrated by two methods. The first was to observe the α-β phase transition of K_2CrO_4 which occurs at 665 °C [15]. The width of the transition temperature range is an upper bound on the width of the temperature gradient across the illuminated section of the sample strip. This was found to be 10 °C, so the temperature variation across the 12 mm width of the sample is ±5 °C.

A second calibration was performed by measuring the lattice parameter of MgO as a function of temperature [16]. This material is cubic and has isotropic thermal expansion characteristics. Diffraction patterns were taken at room temperature and at 100 °C intervals from 500-800 °C. The lattice parameter was determined at each temperature and compared with published data [16]. In this way, the sample temperature can be determined to within ±5 °C.

The sample thickness was determined from the volume absorption coefficient, A_V, defined as

$$A_V = 1 - \exp(-2\mu t / \sin\theta), \qquad (A1)$$

where t is the effective thickness of the sample. It was determined from the ratio of the integrated intensities of several $BaCO_3$ peaks in an aligned sample with those of a so-called "infinitely thick" sample, through which no X-rays penetrate to the heating strip. This procedure was repeated until a sample of suitable effective thickness (typically 50 μm with a μt of about 0.05) was obtained.

APPENDIX B: CONCENTRATION DETERMINATION BY RIETVELD ANALYSIS

We determine the quantitative phase concentrations in the quenched sample by the Rietveld method, which is generally used to refine structural parameters of a powder. The method fits every data point in the diffraction pattern and requires that every phase in the pattern and its structure are known. Several programs are available for doing this analysis. The General Structure Analysis System (GSAS) [17] was used here. There are typically about twenty refinable parameters associated with each phase. To prevent divergence in the refinement process, parameters are gradually added to the refinement. This is done interactively after every few iterations. Figure 3 compares the fitted results (solid line) of a measurement made in helium with the data points (crosses). The lower curve is the difference between the calculated profile and the data points.

The program determines scaling factors, γ_i, which are proportional to the number of unit cells of each phase in the sample. The mole fraction of phase i is given by

$$Y_i = \frac{\gamma_i z_i}{\sum_i \gamma_i z_i}, \qquad (B1)$$

where z_i is the number of molecules per unit cell of phase i. The mass fraction of each phase can be calculated from the γ_i values.

In the absence of preferred orientation, microabsorption, and other minor diffraction effects, the integrated intensity of a reflection (hkl) is given by

$$I_i(hkl) = N_i |F_i(hkl)|^2 M_i(hkl)$$
$$* A_V(hkl) \, LP(hkl) / V_i, \qquad (B2)$$

where subscript i refers to phase i, N_i is the number of illuminated unit cells, F_i the structure factor, M_i the multiplicity factor, A_V the volume absorption factor, LP a geometric correction factor dependent on the reflection angle, and V_i the unit cell volume. The structure factors may differ by orders of magnitude from one phase to the next. The Lorentz-polarization factor is defined as

$$ LP = \frac{1 + \cos^2 2\theta_m \cos^2 2\theta}{4(1 + \cos^2 2\theta_m) \sin^2 \theta \cos \theta}, \qquad (B3) $$

where θ_m is the Bragg angle of the germanium (111) reflection for copper $K\alpha_1$ radiation. It changes by an order of magnitude over the angular range of interest. So, in order to obtain quantitative results, all of these factors must be taken into account.

A diffraction pattern of the unreacted sample was used to determine the absorption correction for the reacted sample. The linear absorption coefficients μ for the carbon and oxygen atoms are very small compared to those of the heavy metal atoms. Thus, μ of the reacted and unreacted sample are about the same.

The Rietveld refinement determined the mole fractions of all the phases, so the corresponding mass fractions can be determined. Assuming no change in the weight of the sample during the high-temperature scans, the mass fraction of the 123 phase can be calculated at any time from the final mass fraction (determined via Rietveld) as

$$ x_{123}(t) = \frac{I_{123}(t)}{I_{123}(\infty)} x_{123}(\infty) \qquad (B4) $$

Determination of Mass Fractions When Rietveld Refinement Cannot be Applied

The integrated intensity of a peak from phase i in a multiphase mixture can also be written as

$$ I_i = x_i I_{i\infty} A_V A_E , \qquad (B5) $$

where x_i is the mass fraction of phase i, $I_{i\infty}$ the intensity of an infinitely thick pure phase i sample, and A_E an attenuation factor accounting for X-ray absorption by the gas. When the integrated intensity is determined in sample 1 which contains only identifiable phases, x_i can be determined by a Rietveld refinement. If sample 2 contains some unidentified phases, this refinement cannot be performed. However, writing Eq. (B5) for the two samples and dividing gives

$$ x_i^{(2)} = \left(\frac{I_i^{(2)}}{I_i^{(1)}} \right) \left(\frac{A_V^{(1)}}{A_V^{(2)}} \right) \left(\frac{A_E^{(1)}}{A_E^{(2)}} \right) x_i^{(1)} \qquad (B6) $$

where the intensities for the infinitely thick pure phase samples cancel out.

There are two sources of inaccuracy in this method. The first is the assumption that most of the mass of the phases present in the oxidizing environment is accounted for by the mass of the metals. The total mass of the sample illuminated by the X-rays can then be assumed the same for all environments (after adjusting for the difference in initial mass via the volume absorption correction). The second and smaller source of error comes from examining only two reflections as opposed to the entire pattern. However, without complete phase information this is a reasonable approximation. In our experiments, the air-processed samples contained unidentified peaks. Equation (B6) was used to determine x_i in air (sample 2) using x_i and I_i in nitrogen (sample 1).

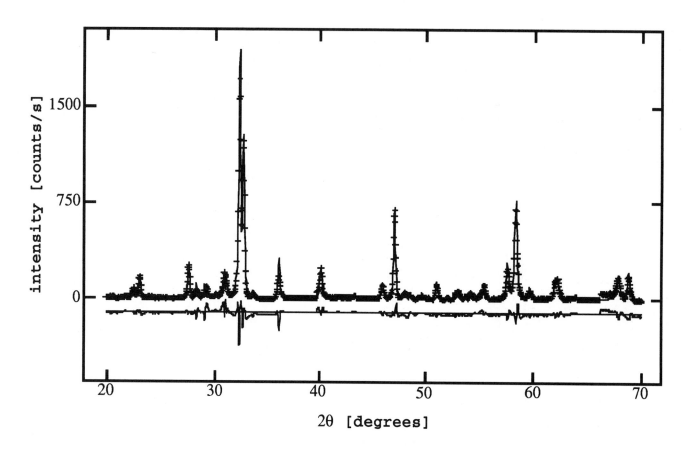

FIG. 3: Rietveld refinement of 123 formation in helium.
Pluses are data points. The curve through the pluses is the fit.
The lower curve is the difference between the fit and the data.

PHASE EQUILIBRIA AND CRYSTAL CHEMISTRY OF HIGH T_c SUPERCONDUCTOR CUPRATES

Winnie Wong-Ng and Lawrence P. Cook ■ Ceramics Division, Natural Institute of Standards and Technology, Gaithersburg, Md. 20899

Knowledge of subsolidus phase compatibilities and melting relations of compounds formed in the Ba-R-Cu-O (R=lanthanides and Y), Tl-Ca-Ba-Cu-O and Sr-Ca-Bi-Cu-O systems is essential for understanding material properties and controlling processing parameters for the cuprate high T_c superconductors discovered in recent years. We report here on the research work concerning phase equilibria and crystal chemistry of these systems being conducted by the Ceramics Division of the National Institute of Standards and Technology to provide this critical information.

Precise phase equilibrium studies of these superconductor systems, in general, were found to involve many difficulties. The variation of oxygen content (which changes the effective oxidation state of Cu), the problems of extensive solid solution formation, compositional changes caused by reaction with containers, and reaction with CO_2 and H_2O, are common to these materials. In the Bi-system, further specific problems are due to the volatility of Bi_2O_3, the difficulty in achieving reproducibility of experimental results, and the persistence of unstable multi-phase assemblages within single phase regions. Also, many Bi-containing phases show incommensurate diffraction due to the modulation of the Bi and/or O positions, rendering the structure determination difficult. Although attainment of equilibrium for each of the stable ternary or quarternary compounds is difficult,

equilibrium can, in general, be more easily achieved further away from the compositions of the stable ternary phases. Therefore, deduction of compatibility joins is often more reliable than delineation of solid solution regions. In the Tl system the toxicity and volatility of thallium oxide, and the disordered nature of compounds with layered structures, pose additional problems.

Experimental investigations of the liquidus relationships have their own set of problems, including difficulty in quenching the liquid phase, difficulties in interpreting X-ray patterns of the complex phase assemblages produced by melt solidification, and the tendency of the liquid to creep out of experimental containers. Very little information on melting is available for the Bi and the Tl systems. In the Y system, only limited data are reported [1-5]. Most of these published liquidus diagrams, however, have been constructed with limited quantitative compositional data on the liquids participating in eutectic or peritectic equilibria. Furthermore, there are apparent conflicts concerning the location and size of primary phase fields of solids and the location of eutectic and peritectic reaction points.

Experimental procedures have been devised to circumvent some of the above-mentioned problems. Research has been conducted to investigate the subsolidus relationships in all three systems, and study of the melting

11

reactions in the Ba-Y-Cu-O system has also been initiated. This paper summarizes some recent results of our phase equilibrium determination efforts in these three systems.

Ba-R-Cu-O SYSTEMS (R — LANTHANIDES AND YTTRIUM)

Low critical current density, flux creep and poor mechanical behavior are a few of the problems which hinder application of the high T_c superconductor material $BaYCu_3O_{6+x}$. The discovery of a variety of lanthanide-substituted compounds of general formula $BaRCu_3O_{6+x}$, which possess similar structures and also exhibit superconducting transition temperatures of $\approx 90K$ [6], prompted us to investigate these alternative materials for possible improvements.

Detailed phase diagram studies for the Ba-Y-Cu-O and Ba-La-Cu-O systems have been reported by Roth et al. [7] and Kilbanow et al. [8], as shown in Figures 1 and 2. A comparison of these diagrams reveals substantial differences in terms of phase formation as well as tie line relationships. The stable valence state and the size of the lanthanide elements play important roles in governing the compound formation in these systems. As one goes across the lanthanide series, the progressive reduction of size, which is known as the lanthanide contraction, allows us to study the effect of R on crystal chemistry, physical properties and chemical properties of the compounds formed. A systematic study of the phase relationships

in the Ba-R-Cu-O systems has been initiated [9-18] in order to characterize the trends of phase formation, solid solution formation, and phase diagram topologies as a function of the size of R.

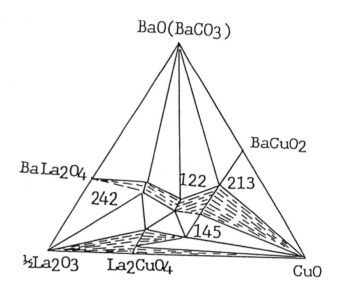

Fig. 2   Phase compatibility diagram at 950°C for the Ba-La-Cu-O system [6].

Phase formation

Selective series of phases of binary and ternary oxides have been prepared in the Ba-R-Cu-O systems for the investigation of phase formation as a relation to the size and the stable valence state of R. Table 1 summarizes our current knowledge. Both literature data and data obtained from our own laboratories have been used to construct this summary [19-24]. Across the top of the table the stable R^{3+} ions including yttrium are listed. The left column illustrates the chemical formulas of the binary oxide phases in the $BaO-R_2O_3$ and R_2O_3-CuO systems and the ternary oxide phases in the $BaO-R_2O_3-CuO$ systems. Inside the table, the symbol 'T' is used to indicate the occurrence of the compound with the respective R. 'O' indicates the absence of such a compound. Further work is still required to complete this table. For details of the phase formation of these compounds, the readers are referred to individual references as listed in the table. The phase formation (including solid solution formation) of the series of "green phase" and "brown phase" compounds and the orthorhombic to tetragonal transformation of the superconducting $Ba_2RCu_3O_{6+x}$ phases will be summarized in the sections below.

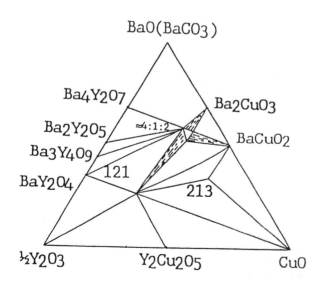

Fig. 1   Phase compatibility diagram at 950°C in air for the Ba-Y-Cu-O system [7].

Table 1. Summary of Phase Formation in the BaO-R_2O_3-CuO Systems (L=La, N=Nd, S=Sm, E=Eu, G=Gd, D=Dy, H=Ho, E'=Er, T=Tm, Y'=Yb and L'=Lu). The symbol 'T': compound forms, 'O': compound does not form.

R^{3+} =	L	N	S	E	G	D	H	Y	E'	T	Y'	L'
R_2O_3-CuO:												
R_2CuO_4 [19]	T	T	T	T	T	O	O	O	O	O	O	O
$R_2Cu_2O_5$ [20,21]	O	O	O	O	O	T	T	T	T	T	T	T
BaO-R_2O_3:												
BaR_2O_4 [15]	T	T	T	T	T	T	T	T	T	O	O	O
$Ba_3R_4O_9$	O	O	T		T	T						
$Ba_2R_2O_5$	O	O										T
$Ba_4R_2O_7$	O	O	O									T
BaO-R_2O_3-CuO:												
$Ba_2RCu_3O_{6+x}$ [22]	T	T	T	T	T	T	T	T	T	T	T	T
BaR_2CuO_5												
green phase [13]	O	O	T	T	T	T	T	T	T	T	T	T
brown phase [13]	T	T	O	O	O	O	O	O	O	O	O	O
$Ba_{3+x}R_{1-x}Cu_{2-z}O_w$ [22]	O	O				T	T					
$BaR_4Cu_5O_{13+x}$ [23,24]	T	O	O	O	O	O	O	O	O	O	O	O
$Ba_3R_3Cu_6O_{14+x}$ [10]	T	T	T	T	O	O	O	O	O	O	O	O
$Ba_{1+x}R_{2-x}Cu_2O_{6-x/2}$	T	O	O	O	O	O	O	O	O	O	O	O

Green Phase" (BaR_2CuO_5) and "Brown Phase" $\underline{Ba_{2+2x}R_{4-2x}Cu_{2-x}O_{10-2x}}$, R=La, Nd).

Under ambient conditions, the well known "green phase," BaR_2CuO_5, has been prepared for R=Sm, Eu, Gd, Dy, Y, Er, Tm, Yb and Lu [13]. This phase, however, does not form with lanthanides of larger size. Among the oxides with a stable R^{3+} valence state, there is a size range within which the substitution of R is possible. For example, the formation of green phases for R=La^{3+} and Nd^{3+} does not take place. The materials formed are brown and are found to have a completely different crystal structure from that of the "green phase." While all green phases are orthorhombic with space group Pbnm(62), Z=4, the "brown phases" tend to form solid solutions of $Ba_{2+2x}R_{4-2x}Cu_{2-x}O_{10-2x}$, with a tetragonal space group of P4/mbm(127). The solid solution range is $0.15 \leq x \leq 0.25$ for the La and $0.0 \leq x \leq 0.1$ for the Nd system.

In the green phase structure, each yttrium ion is surrounded by seven oxygen atoms, as shown in Figure 3. The framework can be considered as built up from distorted monocapped trigonal prisms, RO_7, which share one triangular face forming R_2O_{11} blocks. There is an apparent size limit, bounded by Sm, beyond which the distorted, monocapped, trigonal prism, RO_7, is unstable. The framework of the brown phase La and Nd materials is principally built from edge- and face-sharing BaO_{10} and RO_8 polyhedra [25]. These polyhedra are arranged to provide sufficient space to accommodate the La^{3+} and Nd^{3+} ions. The packing diagram (Figure 4) of the Nd brown phase shows that the Cu atoms have square-planar coordination, with the CuO_4 groups alternating with Ba polyhedra in the xy plane. The Ba, Cu layers are separated by layers of Nd polyhedra perpendicular to the z-axis. All CuO_4 groups are orthogonal to the neighbouring groups. Investigations of the magnetic properties of this material as well as of the solid solution $Ba(La,Nd)_2CuO_5$ are being conducted.

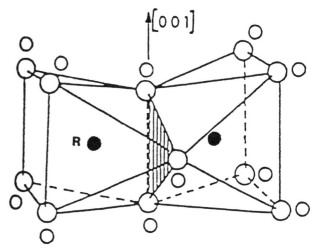

Fig. 3 Polyhedral environment of RO_7 and R_2O_{11} found in the green phase structure.

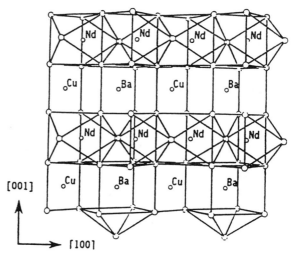

Fig. 4 Packing diagram of the brown phase $BaNd_2CuO_5$.

$Ba_2RCu_3O_{6+x}$ (high T_c superconductors).

The orthorhombic to tetragonal
(nonsuperconducting) structural phase
transformation, which occurs in the high T_c
ceramic superconductors, is of considerable
importance in the processing of these
materials. This phase transformation has
been investigated for six high T_c
superconductors, $Ba_2RCu_3O_{6+x}$, where R = Sm,
Gd, Y, Ho, Er and Nd, and x=0 to 1 [17]. We
have found that the temperature of this
structural phase transition (Figure 5), its
oxygen stoichiometry, and characteristics of
the T_c plateaus (Figure 6) appear to follow a
trend anticipated from the dependence of the
ionic radius of the lanthanide ions. Rare
earth elements with a smaller ionic size can
stabilize the orthorhombic phase to higher
temperatures, or lower oxygen content. Also,
the superconducting temperature is less
sensitive to the oxygen content for materials
with smaller ionic radii [17].

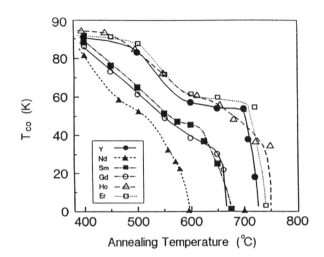

Fig. 6 Superconducting onset temperature,
T_c, vs. phase transition transition
temperature of $Ba_2RCu_3O_{6+x}$.

both the x-ray and neutron diffraction data
indicate a lack of evidence for a long-range
ordering of oxygen in samples quenched around
the "lower" T_c plateau region. For the
larger lanthanide samples (i.e., Nd, Sm and
Gd) the lower T_c plateau may correspond to a
different superlattice resulting from higher
oxygen stoichiometry [17]. Currently,
electron microscopy studies of these samples
are being carried out.

Subsolidus phase relationships

Results of the ternary phase compatibility
diagrams of the systems Ba-R-Cu-O in the
vicinity of the CuO corners, where R = La,
Nd, Sm, Eu, Gd, Y, and Er are shown in Figure
7 [9]. Since exact tie-lines connection
would require detailed lattice parameter
determination, the tie lines connecting the
solid solution series in this report are
schematic only. Proceeding from the La
system, which has the largest ionic size of
R, towards the Er system with a smaller ionic
size, a general trend of phase formation,
solid solution formation, and phase
relationships is found to be correlated with
the ionic size of R.

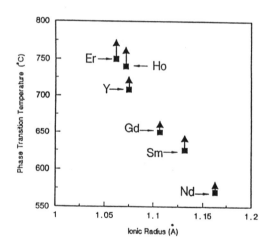

Fig. 5 Orthorhombic to tetragonal phase
transition temperature of $Ba_2RCu_3O_{6+x}$ as a
function of ionic radius.

The transformation from the oxygen-rich
orthorhombic phase to the oxygen-deficient
tetragonal phase appears to involve two
orthorhombic phases. Electron microscopy
studies showed that small areas of most
grains of the Y-compound corresponding to the
second orthorhombic phase had a doubling of
the unit cell dimension along the a-axis,
with a'=2a, indicating absence of oxygen in
every other Cu-O chain along the b-axis.
Based on similar chemical properties, this
type of superlattice can probably be found in
$Ba_2RCu_3O_{6+x}$ compounds with smaller size R.
This is most likely short-range ordering as

Several features of the progressive changes
in the appearance of these ternary diagrams
near the CuO corner are as follows. In
brief, these features are: (1) the La system
has the largest number of ternary compounds
and solid-solution series; this number
decreases as the size of R decreases; (2) the
superconductor material, $Ba_2RCu_3O_{6+x}$, for the
first half of the lanthanide elements (i.e.,
R = La, Nd, Sm, Eu and Gd, which are
gelatively larger in size), exhibits a solid

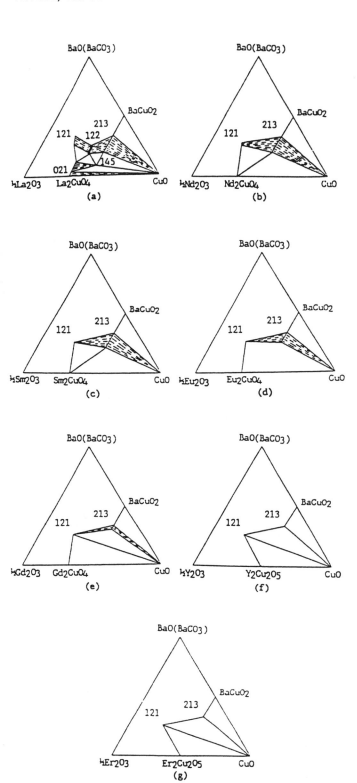

Fig. 7 Subsolidus phase compatibility diagrams of Ba-R-Cu-O near the CuO corner for (a)La, (b)Nd, (c)Sm, (d)Eu, (e)Gd, (f)Y and (g)Er at 950°C in air.

solution $Ba_{2-z}R_{1+z}Cu_3O_{6+x}$ with a range of formation which decreases as the size of R decreases; this solid-solution region terminates at Dy and beyond, where presumably the superconductor phase assumes a point stoichiometry; (3) another feature of these diagrams is illustrated by the tie-line connection from the binary compounds along the $\frac{1}{2}R_2O_3$-CuO edge to the BaR_2CuO_5 phase and the $Ba_2RCu_3O_{6+x}$ series. The R_2CuO_4 compound can only be prepared with the larger size of R, namely, from La to Gd, while the $R_2Cu_2O_5$ phase exists with the smaller size of R. The tie-line connections between the binary $R_2CuO_4/R_2Cu_2O_5$ phases and the BaR_2CuO_5 phase

or the high T_c superconductor compositions appear to reflect the different extent of the $Ba_{2-z}R_{1+z}Cu_3O_{6+x}$ solid solution. When R's are relatively large and the extent of the solid solution line is long, e.g., R = La, Nd and Sm, a compatibility line is found to connect R_2CuO_4 and the tetragonal end member of the $Ba_{2-x}R_{1+x}Cu_3O_{6+x}$ phase. In the systems with R=Eu and Gd, the tie-line connection switches to join the CuO phase and the $BaEu_2CuO_5$ and $BaGd_2CuO_5$ phases, respectively. This trend remains hereafter in the systems with smaller R's.

Melting relationships

Since compositional data describing the liquid are important for constructing a quantitative phase diagram, we have devised a procedure using a combination of experimental methods to circumvent difficulties encountered as mentioned earlier [26]. This procedure includes: (1) calcination and materials handling in special furnace and dry box assemblages, (2) differential thermal analysis/thermogravimetric analysis (DTA/TGA) studies to obtain indication of thermal events, (3) annealing of samples with porous wick materials in order to capture the liquid formed (4) fast quenching of samples in liquid nitrogen- cooled environment for preserving the oxygen stoichiometry of material, (5) powder x-ray characterization of solid phases present, (6) scanning electron microscopy (SEM) studies and x-ray mapping to study the microstructure of the quenched materials, (7) quantitative scanning electron microscopy/electron dispersive x-ray (SEM/EDX) analysis of the composition of melts, and (8) hydrogen reduction to obtain the oxygen content.

This procedure so far has been applied to the studies of the melting of the green phase

BaY_2CuO_5 and the high T_c superconductor phase $Ba_2YCu_3O_{6+x}$ (213). These results show that the peritectic reactions reported in the literature concerning the melting of the green phase and the 213 phase were incorrect. For example, the green phase melts according to the equilibrium: $2BaY_2CuO_5 = Y_2O_3 + BaY_2O_4$ + liquid, as shown in Figure 8, instead of melting to the reported assemblage of Y_2O_3 and liquid. The 213 sample appeared to melt to green phase and two immiscible liquids. We are in the process of confirming this. This results have significant impact on the processing of high T_c superconductors. So far, the melt compositions of these samples have showed only a small content of yttrium (the green phase melt gives the atomic percentage of cations as: Ba 40.9%, Y 2.4% and Cu 56.7%). Oxidation-reduction also plays an important role in melting. For example, when the green phase melts at 1268°C, oxygen is lost, corresponding to a change in bulk CuO_x from $CuO_{1.0}$ to $CuO_{.48}$ (Figure 8). Current work continues to investigate other peritectic and eutectic reactions in the system which will eventually lead to a quantitative liquidus diagram.

THE Tl-Ca-Ba-Cu-O SYSTEM

The long term goal of study of the thallium-based materials is to understand the phase relationships in the quarternary system. The short term goal is to determine the stability of the reported high T_c Ruddelsden-Popper phases, $Tl_xCa_nBa_2Cu_{n+1}O_{2n+3+1.5x}$ (n = number of layers in structure, x= 1 or 2) and the solid solution region of these phases, with emphasis on the phases such as the 2212 (Tl:Ca:Ba:Cu), the 2223, and the 2021 phases. This paper summarizes the results of study of the 2212 phase [27-29].

The high T_c compositions have been investigated in two ways. In the first method the Tl_2O_3 is allowed to diffuse into the starting composition via a vapour route. This is called the buffered vapor transport method [27]. A special apparatus has been designed which employs a buffering technique in which the partial pressure of Tl_2O is held constant by vaporization equilibria in a separate reservoir and the partial pressure of O_2 is held at 1 atm. In the second method, Tl_2O_3 is added to the starting composition and the entire mixture is equilibrated. This is the conventional grinding and annealing/quenching method. Thallium loss can be controlled by the use of tight-fitting containers, and the amount of loss can be accurately determined by weighing before and after the experiments. Full application of the first method requires detailed knowledge of the vapor pressures over thallia containing compositions; this knowledge is as yet incomplete. Consequently

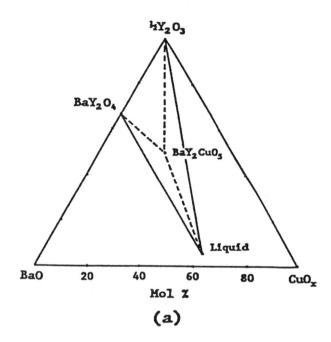

(a)

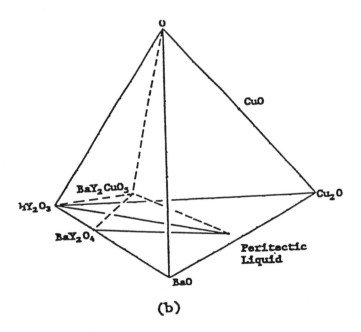

(b)

Fig. 8 Location of composition of invariant melt for the equilibrium: $2BaY_2CuO_5 = Y_2O_3 + BaY_2O_4$ in (a) ternary system and (b) quaternary representation.

only experiments which have utilized the
second method will be discussed here.
Figure 9 illustrates some of the reported
stoichiometries of the Tl-Ca-Ba-Cu-O system
on the Tl-free projected basis.
Compositional relationships can be used from
this diagram for selecting a variety of
starting materials in order to prepare the
high T_c materials. For example, the 2212
compound can be made from a binary mixture of
$BaCuO_2$ and CaO, and the 2223 structure can be
made from a mixture of $BaCuO_2$ and Ca_2CuO_3,
plus thallium oxide.

The processing environment has been found to
have substantial effect on the
crystallization and properties of the high T_c
phases. For example, it has been found that
the reducing atmosphere generated by
stainless steel containers was sufficient to
cause decomposition of the 2212 and 2223
phases. Such decomposition with progressive
grinding and annealing had initially led to
the posulated metastability of these phases
[27]. The 2212 phase was subsequently shown
to be stable in both air and oxygen
atmospheres by performing annealing
experiments in Au capsules [28]. A diagram
of the experimental setup for these
experiments is shown in Figure 10.

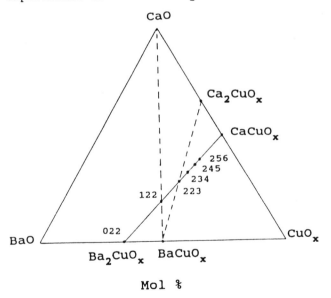

Fig. 9 Stoichiometries of the Tl-Ca-Ba-Cu-O
system on the Tl free projected basis.

A significant effect was observed during the
studies of the T_c of these 2212 phases. The
T_c values varied extensively, not with the
annealing times as expected, but with the
amount of Tl_2O_3 lost during the annealings.
The maximum observed T_c (105K) occurred for

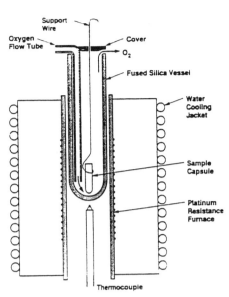

Fig. 10 Experimental setup for the oxygen annealing
experiments.

compositions with significantly less Tl_2O_3
content than the ideal 2122 stoichiometry at
16.7 mol%. In order to investigate the
possibility of solid solution formation of
the 2212 phase, experiments were conducted
using mixtures of precalcined 0122
composition and Tl_2O_3 at 700-900°C in tightly
crimped Au capsules under an atmosphere of
flowing O_2. Formation of the 2122 phase was
observed between 5-25 mol% Tl_2O_3.
Interpretation of the total collection of x-
ray patterns with regard to the presence of
the 2122 phase is summarized in Figure 11.
The lack of evidence for substantial amounts
of additional phases over a wide range of
composition suggested possible solid solution
formation. A region of "nearly single phase
2122" is centered around the ideal 2122
composition, and lies slightly off the
$CaBa_2Cu_2O_x$ - Tl_2O_3 join. It extends at 860°C
from 9-13% to 20-25 mol% Tl_2O_3. This region
is not present at 900°C nor at 820 °C and
below. As would be expected, the 2122 phase
is the major constituent in the T-X
(temperature-composition) area surrounding
the nearly single phase region, but by
contrast with earlier experiments, Tl loss
was minimized in these experiments, which
were of short duration. Yet, in agreement
with initial observations on materials from
longer experiments, there is a correlation
between Tl content and T_c, with the maximum
T_c of 109K occurring for Tl_2O_3 contents of
approximately 5 mol%, again, substantially
less than the ideal 2122 stoichiometry.

Undoubtedly, a contributing factor in the
phase formation or the progression between

phases is the oxidation state of the Tl oxide. It is also conceivable that under conditions of longer term equilibration, there could be some tendency for substitution of Tl by Ca and Ba, and vica versa. Detailed

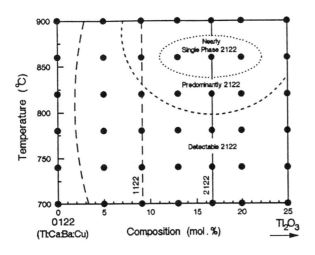

Fig. 11 Temperature-composition plot showing experiments completed under an oxygen atmosphere, with extent of 2122 phase formation.

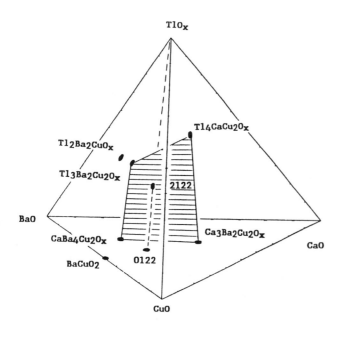

Fig. 12 Proposed solid solution plane for $Tl_2CaBa_2Cu_2O_8$.

physical models for the high Tc mechanism may depend upon these stoichiometry/T_c relations. Various models have been proposed in literature to explain the variations of T_c in the Ruddelsden-Popper series, but most are limited by a lack of knowledge of stoichiometric variations [30,31,32]. At this stage, we are still without proven crystal-chemical models for this solid solution variation in the Tl-based materials. However, assuming no cation vacancies, hypothetical end members for these stoichiometric variations can be derived as shown in Figure 12, which illustrates the proposed solid solution plane for the 2122 phase. Work is continuing on the relation of T_c to the proposed solid solution plane.

THE Sr-Ca-Bi-Cu-O SYSTEM

The phase equilibrium studies in the Sr-Bi-Ca-Cu-O system have been a major undertaking by Roth and coworkers [33-38]. The compatibility relations in the quarternary system are still under investigation, but considerable work has been performed on various binaries and ternaries. Here, we summarize the results pertaining to the systems Bi_2O_3-SrO-CaO[33,38], Bi_2O_3-CaO-CuO[34,37], SrO-CaO-CuO[36], and Bi_2O_3-SrO-CuO[33,34].

SrO-CaO-CuO

A partial isothermal diagram at 950°C is shown in Figure 13. This ternary system has three solid solution series which extend from

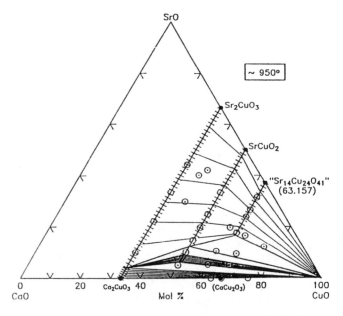

Fig. 13 Phase diagram for the system SrO-CaO-CuO at 950°C.

the SrO-CuO join toward the CaO-CuO join, and at least one ternary phase is present. Only one solid solution series, the 2:1 [$(Sr,Ca)_2Cu_1O_3$] extends all the way across to a pure Ca end member. The compounds Ca_2CuO_3 and Sr_2CuO_3 are essentially isostructural and form a complete series of solid solutions. The 14:24 series of solid solutions extends from the compound $Sr_{14}Cu_{24}O_{41}$, which was a new phase discovered at NIST, to approximately the composition $Sr_7Ca_7Cu_{24}O_{41}$. The 1:1 solid solution series is based on the structure of $SrCuO_2$, with Ca^{2+} substituting for Sr^{2+} up to 75% Ca. Although at 950°C, the pure $CaCuO_2$ end member is less stable than the assemblage Ca_2CuO_3 + CuO, at higher temperatures a nonstoichiometric phase $Ca_{1-x}CuO_2$ is formed, and below 800°C, $CaCuO_2$ appears to have a field of stability [39]. The system has at least one ternary phase with a small homogeneity range, $(Sr_xCa_{1-x})CuO_2$, x≈0.12 to 0.15. The structure of this compound can be considered as an end member of the homologous series $(Tl,Bi)_2(Ba,Sr)_2Ca_{n-1}Cu_nO_{4+2n}$ [40].

$SrO-Bi_2O_3-CuO$

Figure 14 is a composite diagram of subsolidus data. The main features of this diagram are the presence of four ternary phases and several series of solid solution. The first phase, $Sr_2Bi_2CuO_6$ (Sr:Bi:Cu 221, or Sr:Bi:Ca:Cu 2201), was found to be monoclinic, with a small homogeneity range. This phase always contained a small amount of the $Sr_{14}Cu_{24}O_{41}$ phase and did not exhibit superconductivity. It is not known if this is a true solid solution region or a collection of smaller regions in which several structurally related phases are stable. The commonly referred to 20 K superconducting '2201' phase was determined to be a Raveau-type solid solution but with composition distinctly different from 2201. It is a tetragonal solid solution corresponding approximately to the formula $Sr_{1.8-x}Bi_{2.2+x}Cu_{1\pm x/2}O_z$. A two-phase region is found to exist between the 2201 phase and this Raveau-type phase. The Raveau phase has a large stable compositional region of solid solution with a gross deficiency in SrO, whereas the 2201 composition has only a slight deviation from ideal stoichiometry and is probably always less than about one mole % deficient in CuO.

There is a two-phase region involving CuO and the rhombohedral solid solution on the SrO-Bi_2O_3 edge. CuO is also in equilibrium with a range of saturated compositions comprising

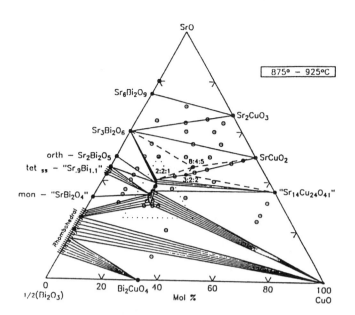

Fig. 14 Phase diagram for the system $SrO-\frac{1}{2}Bi_2O_3-CuO$.

the Raveau-type solid solution region. The compound $Sr_{14}Cu_{24}O_{41}$ is in equilibrium with all three of the ternary phases related to the structurally homologous series, namely, the 221, 322, the Raveau solid solution, but not with the structurally dissimilar phase 8:4:5.

$CaO-Bi_2O_3-CuO$

The predominant feature of the $CaO-Bi_2O_3-CuO$ diagram is the absence of any ternary phase, as shown in Figure 15, an isothermal section

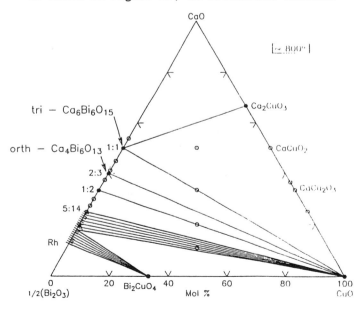

Fig. 15 Phase diagram for the system $CaO-\frac{1}{2}Bi_2O_3-CuO$ at 800°C.

at ~800°C. CuO is found to be in equilibrium with all the CaO-½Bi₂O₃ binary phases. The compound Bi_2CuO_4 is in equilibrium only with the higher Bi_2O_3 portion of the rhombohedral solid solution. Along the CaO-Bi₂O₃ join, two new phases have been identified, the triclinc $Ca_6Bi_6O_{15}$ phase and the orthorhombic $Ca_4Bi_6O_{13}$ phase. Compositions of these phases have been revised from previous references, where they were incorrectly indicated as 7:6 and 7:10, respectively [41].

SrO-CaO-½Bi2O3

A large number of solid solution series appear in this system at 800-900°C due to the substitution of the smaller Ca^{+2} for the larger Sr^{+2}, as shown in Figure 16. The phases of composition SrO:1/2Bi₂O₃ of 3:1, 3:2, 1:1, 9:10, 1:2 accept CaO to form solid solutions to different extents. The rhombohedral solid solution (Rh$_{ss}$) is complete across the entire range of SrO-CaO ratios. No SrO has been found in solid solution in either $Ca_4Bi_6O_{13}$ or $Ca_2Bi_2O_5$. Two ternary phases (solid solutions) have been discovered. The one with higher Sr content is a solid solution with a general formula $A_4Bi_2O_7$ (i.e., $Sr_3CaBi_2O_7$), and the other one is of general formula $A_2Bi_2O_3$ (i.e., $CaSrBi_2O_5$). These both have monoclinic symmetry.

Crystal Structures

During the course of investigation of these systems, a substantial number of new phases have been discovered or confirmed. Single crystals of these phases have been prepared and their structures subsequently have been determined by using the single crystal x-ray technique or in combination with the Rietveld refinement technigue. Since x-ray powder diffraction patterns are essential for material identification, some of these patterns have also been prepared. Examples of new phases whose structures have been determined are: $Ca_{1-x}CuO_2$ [39], $Sr_{14}Cu_{24}O_{41}$ [41], $(Sr_{0.16}Ca_{0.84})CuO_2$ [42], $Ca_6Bi_6O_{15}$ [43], $Ca_4Bi_6O_{13}$ [44], $CaBi_2O_4$ [45], $Sr_2Bi_2CuO_6$ [46], $Sr_2Bi_2O_5$ [47] and $Ca_2Bi_2O_5$ [48]. Among these phases, none of which is superconducting, three are selected to be discussed either because of the special bonding features or because of relation to the parent structure of the superconducting compound series.

$(A_{1-x}A_x')_{14}Cu_{n4}O_{41}$

The structure of this new series of compounds, where A = Sr, Ca or Ba, and A' = La, Y or Bi, have been determined [40] to have a face-centered orthorhombic subcell, with a=11.3 Å, b=13.0 Å, and c=3.9 Å. A packing diagram of the structure is illustrated in Figure 17. Superstructure is observed in crystals, leading to a seven-fold increase of the c-axis and a change in symmetry to space group Cccm. This new structure contains both linear Cu-O chains and Cu-O planes. Edge sharing in the chains as well as in the planes leads to short Cu-Cu contacts. Edge sharing in the planes also produces a zig-zag arrangement.

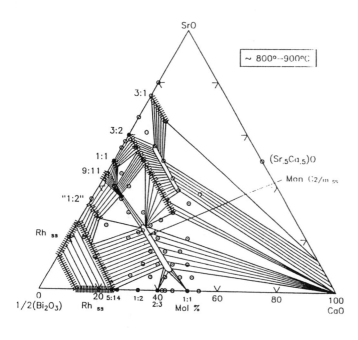

Fig. 16 Phase diagram for the system SrO-½Bi₂O₃-CaO at 800-900°C.

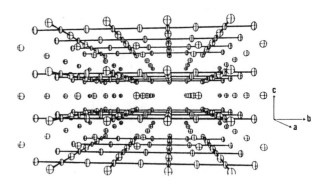

Fig. 17 Extended view of $(Ca_{0.86}Sr_{0.14})CuO_2$. Only the Cu-O bonds are drawn.

$(Ca_{0.86}Sr_{0.14})CuO_2$

The crystals are tetragonal with space group P4/mmm, a=3.8611(2), c=3.1995(2) Å. It is a simple perovskite built from square-planar CuO_2 sheets that sandwich Ca and Sr ions (Figure 18). It is regarded as the parent structure of $A_2B_2Ca_{n-1}Cu_nO_{4+2n}$ for large n in the case of the Tl-Ba-Ca-Cu-O and Bi-Sr-Ca-Cu-O superconductor systems, where additional $CaCuO_2$ layers are added. Taking away the thallim oxide and bismuth oxide layers results in a regular coordination for Ca, Sr and Cu which leads to high symmetry and a small unit cell.

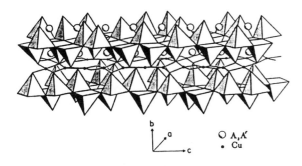

Fig. 18 Structure of $(A_{1-x}A'_x)_{14}Cu_{24}O_{41}$, side view of 2 Cu-O planes linked with the Cu-O chains. Cu atoms are indicated by small filled circles, the oxygen atons are at the vertex. A = Sr, Ca or Ba, and A' = La,Y or Bi.

$Sr_2Bi_2O_5$

This structure has ordered cation arrangements and resulting lower symmetry. The compound crystallizes in the orthorhombic space group Pnma, a=14.261(3), b=6.160(2), and c=7.642(3) Å. This is one example demonstrating a new low bismuth coordination environment. All bismuth atoms are in an unusual threefold coordination with oxygen. Figure 19 shows a chain of 3-coordinated bismuth atoms in $Sr_2Bi_2O_5$ oriented along the c-axis. The oxygen atoms alternate with vacancies along the c-axial direction. The

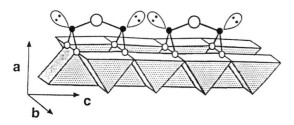

Fig. 19 Structure of $Sr_2Bi_2O_5$.

Bi 6s lone pair of electrons can be envisioned as being directed toward the vacant site, as is represented in schematic form. The Sr atoms lie in the triangular prismatic coordination polyhedra below the bismuth atoms in the diagram.

SUMMARY

The current research results of the crystal chemistry and phase equlibrium studies in the Ba-R-Cu-O (R=lanthanides), Sr-Ca-Bi-Cu-O and Tl-Ca-Ba-Cu-O systems being conducted at NIST are highlighted in this paper. The study of the Ba-R-Cu-O system revealed the trend of phase formation and subsolidus phase diagrams as a function of the size of the ion R. Work is continuing on the liquidus relationships, and a set of experimental procedures has been designed to study quantitative melting relationships. Research in the Ba-Ca-Tl-Cu-O system included the study of the stability and solid solution region of the superconducting 2122 (Ba:Ca:Tl:Cu) phase. A correlation between solid solution stoichiometric variations and variations in T_c in the 2122 phase has been confirmed. Investigation of the ternary and binary diagrams of the Sr-Bi-Ca-Cu-O systems showed complicated phase relationships. Extensive solid solutions existed in all ternary systems. Many new phases have been discovered and their crystal structure being determined. Currently, the quarternary system is still under investigation.

REFERENCES

1. Aselage, T. and Keefer, K., J. Mater. Res. **3** [6], 1279 (1988)

2. Lay, K.W. and Renlund, G.M., Tech. Rep. No. 89CRD096, General Motors Res. & Devel. Center, (1989).

3. Roth, R.S., Davis, K.L., and Dennis, J.R., Adv. Ceram. Mater. **2** [3B], 303 (1987).

4. Maeda, M., Kadoi, M. and Ideda, T., Jap. J. Appl. Phys. **28** [8], 1417 (1989).

5. Nevriva, N., Holba, P., Durcik, S., Zemanova, D., Pollert, E. and Triska, A., Physica **C157**, 334, (1989).

6. Le Page, Y., Siegrist, T., Sunshine, S.A., Schneemeyer, L.A., Murphy, D.W., Zahurak, S.M., Waszczak, J.V., McKinnon, W.R., Tarascom, J.M., Hull, G.M., and Greene, L.H., Phys. Rev. **B63**, 3617 (1987).

7. Roth, R.S., Rawn, C., Whitler, J., and Beech, F., Ceramic Superconductors II, edited by M.F. Man, Am. Ceram. Soc., 13 (1988).

8. Kilbanow, D., Sujata, K., and Mason, T.O., J. Amer. Ceram. Soc. 71(5), C267 (1988).

9. Wong-Ng, W., Paretzkin, B. and Fuller, Jr., E.R., J. of Solid State Chem. 85, 117 1990.

10. Wong-Ng, W., Paretzkin, B., and Chiang, C.K., Powd. Diff. 5, No.1, 26 (1990).

11. Wong-Ng, W., Cook, L.P., Chiang, C.K., Swartzendruber, L.J., Bennett, L.H., Blendell, J.E., and Minor, D., J. Mater. Res. 3 [5], 832 (1988).

12. Wong-Ng, W., Cook, L.P., Chiang, C.K., Swartzendruber, L.J., and Bennett, L.H., Ceramic Superconductors II, edited by M.F. Yan, Am. Ceram. Soc., 27 (1988).

13. Wong-Ng, W., Kuchinski, M.A., McMurdie, H.F. and Paretzkin, B., Powd. Diff. 4 [1], 1 (1989).

14. Wong-Ng, W., and Paretzkin, B., Powd. Diff., in print (1991).

15. Wong-Ng, W., Cook, L.P., Chiang, C.K., Vaudin, M.D., Kaiser, D.L., Beech, F., Swartzendruber, L.J., Bennett, L.H. and Fuller, Jr., E.R., High Temperature Superconducting Compounds: Processing & Related Properties, edited by S. H. Whang and A. DasGupta, Am. Ceram. Soc., 553 (1989).

16. Blendell, J.E., Wong-Ng, W., Chiang, C.K., Shull, R.D., and Fuller, Jr., E.R., High Temperature Superconducting Compounds: Processing & Related Properties, edited by S. H. Whang and A. DasGupta, Am. Ceram. Soc., 193 (1989).

17. Wong-Ng, W., Cook, L.P., Chiang, C.K., Vaudin, M.D. and Bennett, L.H., J. Ceram. Soc., submitted to J. Am. Ceram. Soc. (1991).

18. Wong-Ng, W., Cook, L.P., Paretzkin, B., and Hill, M.D., MRS symposium Proceedings: High-Temperature Superconductors: Fundamental Properties and Novel Materials Processing, 169, 81 (1990).

19. Muller-Buschbaum, von Hk., and Wollschlager, W., Z. Anorg. Allg. Chem. 414, 76 (1975).

20. Freund H. -R., and Muller-Buschbaum, von

Hk., Z. Naturforsch, 32B, 609, (1977).

21. Lambert, E., JCPDS Grant-in-Aid Report, (1981) and (1982).

22. Herman, F., Phys. Rev. B 37, 2309 (1988).

23. Michel, C., Er-Rakho, L. and Raveau, B., Mater. Res. Bull. 20, 667 (1985).

24. Michel, C., Er-Rakho, L., Hervieu, M., Pannetier, J., and Raveau, B., J. Solid State Chem. 68, 143 (1987).

25. Stalick, J. and Wong-Ng, W., Mater. Let. 9[10], 401 (1990).

26. Wong-Ng, W. and Cook, L.P., Ceramic Transactions 18, Superconductivity in Ceramic Superconductors II, edited by K.M. Nair, U. Balachandran, Y-M. Chiang and A. Bhalla, published by the Amer. Ceram. Soc., Westerville, Ohio, 73-84 (1991).

27. Cook, L.P., Wong-Ng, W., Chiang, C.K., and Bennett, L.H., Proceeding of the Superconductor Symposium of the 1st International Ceramic Science and Technology Congress, Anaheim, California, Am. Ceram. Soc., 329 (1989).

28. Cook, L.P., Chiang, C.K., Wong-Ng, W., Swartzendruber, L.J., and Bennett, L.H., MRS symposium Proceedings: High-Temperature Superconductors: Fundamental Properties and Novel Materials Processing, 169, 137, (1990).

29. Cook, L.P., Wong-Ng, W., Chiang, C.K., and Bennett, L.H., Ceramic Transaction 18, Superconductivity in Ceramic Superconductors II, edited by K.M. Nair, U. Balachandran, Y-M. Chiang and A. Bhalla, published by the Amer. Ceram. Soc., Westerville, Ohio, 65-71 (1991).

30. Casella, R.C., Solid State Commun. 70 [1], 75 (1989).

31. Casella, R.S., Solid State Commun. 74 [5] 377 (1991).

32. Casella, R.C., Nuovo Cimento 10D [12] 1439 (1988).

33. Roth, R.S., Rawn, C.J., Burton, B.P., and Beech, F., J. Res. Natl. Inst. Stand. Technol. 95, 291 (1990).

34. Roth, R.S., Rawn, C.J., and Burton, B.P., Ceramic Transactions 13, Superconductivity and Ceramic Superconductors. Edited by K.M.

Nair and E.A. Giess, published by Am. Ceram. Soc., Westerville, OH, 23, (1990).

35. Roth, R.S., Hwang, N.M., Rawn, C.J., Burton, B.P., and Ritter, J.J., J. Am. Ceram. Soc. **74**, (1991), in print.

36. Roth, R.S., Rawn, C.J., Ritter, J.J., and Burton, B.P., J. Am. Ceram. Soc. **72**[8] 1545 (1989).

37. Burton, B.P., Rawn, C.J., Roth, J.S. and Hwang, N.M., J. Res. Natl. Inst. Stand. Technol., in press (1991).

38. Roth, R.S. and Rawn, C.J., J. Res. Natl. Inst. Stand. Technol., in press (1991).

39. Siegriest, T., Roth, R.S., Rawn, C.J. and Ritter, J.J., Chem. Mater. **2**, 192 (1990).

40. Siegrist, T., Schneemeyer, L.F., Sunshine, S.A., Waszczak, J.V., and Roth, R.S., J. Mater. Res. Bull. **23**, 1429 (1988).

41. Conflant, P., Boivin, J.C., and Thomas, D., J. solid State Chem. **18**, 133 (1976).

42. Siegrist, T., Zahurak, S.M., Murphy, D.W., and Roth, R.S., Nature (London) **334**, 231 (1988).

43. Siegrist, T., Roth, R.S., and Rawn, C.J., in preparation, (1991).

44. Parise, B., Torardi, C.C., Whangbo, M.H., Rawn, C.J., Roth, R.S. and Burton, B.P., Chem. Mater. **2**, 454 (1990).

45. Wong-Ng, W., Roth, R.S., Rawn, C.J., and Burton, B.P., in preparation, (1991).

46. Roth, R.S., Rawn, C.J., and Bendersky, L.A., J. Mater. Res. **5**[1] 46 (1990).

47. Torardi, C.C., Parise, J.B., Santoro, A., Rawn, C.J., Roth, R.S. and Burton, B.P., J. Solid State Chem. **93**, in print (1991).

48. Parise, J.B., Torardi, C.C., Rawn, C.J., Roth, R.S., Burton, B.P., and Santoro, A., Chem. Mater., in press. (1991).

STUDIES OF SUBSTITUTION IN $RBa_2Cu_3O_y$

Lawrence Suchow ■ Department of Chemical Engineering, Chemistry, and Environmental Science, New Jersey Institute of Technology, Newark, New Jersey 07102

0.33-33% of the Cu in superconducting $YBa_2Cu_3O_y$ has been replaced by Li (i.e., x = 0.01-1 in single-phase or nominal $YBa_2Cu_{3-x}Li_xO_y$, x-ray diffraction powder patterns remain the same as $YBa_2Cu_3O_y$, with identical patterns up to about 17% substitution (i.e., x = 0.5). At higher percentages an additional phase appears. Electrical conductivity measurements indicate a small elevation of T_c at low Li content. Starting at about 5% Li (x = 0.15), T_c declines progressively and its width increases as x is raised. Attempts to substitute Bi for Nd in orthorhombic $NdBa_2Cu_3O_y$ prepared in air or oxygen at about 950°C led instead to formation of Ba_2NdBiO_6, a new cubic compound with a = 0.8703 nm. The possibility was then explored of preparing superconducting $(Nd_{1-x}Bi_x)Ba_2Cu_3O_y$ by first forming the tetragonal phase at 880-950°C in nitrogen or argon followed by reheating in oxygen or air at 250-500°C in order to insert the additional oxygen required to yield the orthorhombic form while avoiding oxidation of Bi^{3+} to Bi^{5+}. X-ray diffraction studies, electrical conductivity measurements, and thermogravimetric analysis of products indicate that Bi does not enter the $NdBa_2Cu_3O_y$ lattice in either the tetragonal or the orthorhombic phase. Ba_2NdBiO_6 clearly forms on reheating in oxygen or air even at low temperatures, and evidence is presented that a poorly crystallized oxygen-deficient form of this compound is already present prior to the reheating.

In 1986, Bednorz and Müller [1] reported the discovery of the superconductor $(La,Ba)_2CuO_{4-\delta}$, with a T_c of about 30K. Subsequent feverish worldwide activity on "high-T_c" superconductors resulted in preparation of a number of additional compounds, including those with the general formula $RBa_2Cu_3O_y$ (with y≈7), which exhibit T_c values of about 92K [2].

At New Jersey Institute of Technology in the summer of 1987, a solid state chemistry professor (the author of this paper) three electrical engineering professors (Roy H. Cornely, Walter F. Kosonocky, and Kenneth S. Sohn), and their respective students joined forces to study the new superconductors. Two preparative approaches were taken:

(1) Bulk material was synthesized via solid state reaction.

(2) Thin films were obtained by sputtering from bulk $RBa_2Cu_3O_y$ targets.

Much of the bulk preparation work dealt with studies of substitution of various ions into $RBa_2Cu_3O_y$, and two such investigations will be summarized in the present paper.

STUDIES OF LITHIUM SUBSTITUTION [3]

There have been many investigations of the effects of various impurities on the properties of $YBa_2Cu_3O_y$. Yan, Rhodes, and Gallagher [4] included lithium among their dopants and chose to prepare $YBa_2Cu_3O_y$ first and then to react this with a lithium compound (indicated to be either the oxide or hydroxide). It was assumed that a small ion like Li^+ would replace Cu^{2+}. However, for this to occur some of the Cu would have to be forced from its sites in the prereacted $YBa_2Cu_3O_y$. The more heavily doped samples were found to contain small amounts of Y_2BaCuO_5. Obviously, then, the $YBa_2Cu_3O_y$:Li preparation could not be stoichiometric. Regardless, it was found by a.c. susceptibility measurements that Li depresses T_c. Dou et al. [5] also saw a depression in T_c as measured electrically. Both groups found little, if any, change in lattice parameters.

Wang et al. [6] studied the effects of addition of $LiNO_3$, Li_2CO_3, LiF, and $BaLiF_3$ fluxes on the sintering of $YBa_2Cu_3O_y$ and on the resulting microstructures and superconductive characteristics. Most of their studies involved addition of the fluxes to $YBa_2Cu_3O_y$ but they also attempted preparations of $YBa_{2-x}Cu_{3-x}O_y·xBaLiF_3$, where they hoped that Li would replace Cu; however, BaF_2 was precipitated.

The work described in the current paper was initiated prior to our learning of these earlier studies, but it was continued because the approach adopted was different. We were interested in determining whether partial substitution of Li^+ for Cu^{2+} would stabilize Cu^{3+} ions, much as Li^+ doped into NiO stabilizes Ni^{3+}. At the time of writing of this paper, there is a controversy as to whether $YBa_2Cu_3O_y$ contains Cu^{3+} at all or whether the extra positive charge (or hole) is instead on an oxygen anion, resulting in O^- rather than O^{2-}. These two models are different essentially only in the electron density distribution along a Cu-O bond, and for the purpose of this paper, Cu^{3+} presence will be assumed. As stated above, it is normally considered that a small ion like Li^+ will replace only Cu^{2+} in $YBa_2Cu_3O_y$, and our stoichiometry is based on this assumption, but there is some possibility that Li^+ could also simultaneously substitute for Y^{3+}. Relevant Shannon [7] 'IR' radii are given in the following table:

Ion	Coordination Number (CN)	r(nm)
Cu^{2+}	4	0.057
Cu^{2+}	4 (Square)	0.057
Cu^{2+}	5	0.065
Cu^{2+}	6	0.073
Cu^{3+}(Low-spin)	6	0.054
Y^{3+}	6	0.0900
Y^{3+}	8	0.1019
Li^+	4	0.0590
Li^+	6	0.076
Li^+	8	0.092

It is therefore seen that Li^+ is 3.5% larger than Cu^{2+} for CN 4 and 4.1% larger than Cu^{2+} for CN 6, while Li^+ is 9.7% smaller than Y^{3+} for CN 8. If one invokes the rough 15% rule, it does appear possible that some Li^+ ions could go into the distorted cubic Y^{3+} sites while others simultaneously enter the clearly preferred copper sites.

Experimental (Li)

Preparations were based on the general formula $YBa_2Cu_{3-x}Li_xO_y$, with x-values from 0 to 1. In the series to be emphasized in this paper ("Set 1") the preparations were made by first thoroughly grinding together with an agate mortar and pestle Y_2O_3 (99.99%, Research Chemicals), $BaCO_3$ (Mallinckrodt Analytical Reagent), CuO (Mallinckrodt Analytical Reagent), and Li_2CO_3 (99.8%, CERAC) in the appropriate proportions to yield the desired compositions. The ground mixtures were transferred into open porcelain crucibles which were placed in a Harper muffle furnace at room temperature. Furnace power was then turned on and the temperature raised to the 910-925°C range. The purpose of this was to allow the Li_2CO_3 to react slowly

while still in the solid state rather than causing it to melt rapidly, in which case it might escape to and react with the crucible. The maximum temperature range was deliberately chosen to be on the low side of the usual range employed because a lithium compound is expected to act as a flux during the reaction and because Li^+ reduces the melting point of the final product. The mixtures were held at 910-925°C for 16 hours and were then reground, reheated for 16 hours at the same temperature, and cooled in the furnace, whose large size and heat capacity result in slow cooling. With power off overnight, the furnace is found to be at about 100°C in the morning.

For comparison, some additional Li_2CO_3-containing mixtures were prepared as above but with maximum temperature of 950°C; or as above but with sudden heating of reaction mixtures in a furnace already at the reaction temperature of 910-925 or 950°C.

A full series ("Set 2") was also prepared using an aqueous $LiNO_3$ solution as the source of Li^+; these were actually prepared by dissolving appropriate amounts of Li_2CO_3 in excess nitric acid and adjusting all to the same volume with H_2O. Where no Li^+ was added (i.e., x=0), water alone was used for better comparison. After thorough mixing, the liquid was removed by evaporation in an oven at 110°C. The mixtures were then treated as in the first Li_2CO_3 set described above (i.e., ground mixtures placed in furnace at room temperature and power then turned on to attain 910-925°C). The purpose of this series was to determine whether $LiNO_3$ coming out of solution might coat other reactant particles better and result in more uniform distribution of Li^+ in the final product.

In all cases, the products after the second heating were reground and shaped into 1/2-inch-diameter discs in a cylindrical die in a Carver hydraulic press with the powders subjected to a pressure of 0.14-0.28 GPa (2.0-4.0 x 10^4 lbs/in^2). The pellets were then placed in the furnace at room temperature, sintered for 24 hours at 910-925°C (regardless of original preparation temperature because the Li-containing materials melt at 950°C), cooled in the furnace at its natural slow rate to 400°C, held at that temperature for 16 hours, cooled slowly in the furnace to 100°C, and then removed from the furnace. Finally, the pellets were annealed for two hours at 420°C in flowing oxygen in a one-end-closed combustion tube in the muffle furnace. The tube was removed from the furnace with continuing oxygen flow until room temperature was reached.

Results and Discussion (Li)

X-ray diffraction study of the $YBa_2Cu_3O_y$ preparation (i.e., x=0 in $YBa_2Cu_{3-x}Li_xO_y$) indicates that the lattice constants and intensities are essentially the same as those widely reported in the literature, for example by Wong-Ng et al.[8], for the desired orthohombic form. Our lattice constants are a=0.383, b=0.389, c=1.164 nm.

In the "Set 1" (Li_2CO_3, 910-925°C) and "Set 2" ($LiNO_3$, 910-925°C) preparations described above, all x-ray pattern positions and intensities are the same from x=0 through 0.30, but the patterns of preparations with Li_2CO_3 are stronger and with narrower lines than of those with $LiNO_3$. At x=0.50, the orthorhombic compound pattern remains unchanged in both sets but weak additional lines are seen in the $LiNO_3$ preparation and the Li_2CO_3 preparation still has a sharper pattern. At x=1, the orthorhombic compound pattern of the $LiNO_3$ preparation is all or mostly gone and the extra-line phase(s) seen in the x=0.50 sample has become the major constituent. However, with Li_2CO_3 at x=1, the orthorhombic compound pattern is still rather strongly present but additional lines, apparently like those in the x=1 preparation from $LiNO_3$, are also in evidence.

One must try to understand why lattice parameters and intensities remain remarkably the same over so wide a range of composition. This has been observed not only in the current study but also in References [4] and [5]. The possibility that Li^+ does not actually enter the lattice must be considered but it is likely that impurity lines would be observed in preparations with less Li^+ if this were the case because the stoichiometry would be off. Also, changes in T_c are seen. It is conceivable from arguments made above that small amounts of Li^+ replace Y^{3+} as larger amounts simultaneously replace Cu^{2+}, with the result that the two counterbalance each other. Whether the intensities would then also be expected to remain the same would have to be determined by structure factor calculations.

The results of the electrical measurements on "Set 1" are given in Table I and in Figures 1, 2, and 3. Some typical resistance vs. temperature results are shown in Figure 1. When resistance increases with temperature, the material is usually considered to be metallic; when the opposite is the case, it is usually considered to be a semiconductor. Figure 1 and Table I, by these assumptions, indicate metallic behavior above T_c in some (though with varying R vs. T slopes), but others, as with x=0.30 and 0.50, exhibit a resistance reversal, which might be interpreted as a gradual metal-semiconductor transition.

Table I and Figures 2 and 3 indicate (despite some point scatter) a small increase in T_c from x=0 (0% Li) through 0.09 (3% Li) which could be due to stabilization of Cu^{3+} by Li^+ before its negative effects are felt, or could perhaps be due to more effective reaction as a result of the flux-like behavior of the Li compound. Starting at x=0.15 (5% Li), T_c falls and the transition widens as x is increased. Apparently when 5% or more of the Cu^{2+} is replaced by Li^+ the electron transport network is no longer as continuous as required for optimum superconductivity. The decline in T_c which we observe is similar to that seen by Yan et al.[4] but we find much less sensitivity to Li content. However, as pointed out above, all of their preparations had to be nonstoichiometric and probably multi-phase because they were from prereacted $YBa_2Cu_3O_y$ and excess Li^+. Also, their measurements were of magnetic susceptibility rather than electrical conductivity.

Regarding the other preparations described above ($LiNO_3$ 910-925°C; and Li_2CO_3, 950°C), results with Li_2CO_3 were generally much like those described in greater detail for "Set 1" but with somewhat erratic character. It is not entirely clear whether preparation temperature over this narrow range is a very important factor but the procedure employed for "Set 1" is considered to be most reliable. As stated above, x-ray diffraction indicated deterioration in structure of the $LiNO_3$ preparations ("Set 2") at high Li content and resistance measurements are consistent with this. T_c measurements parallel those of the Li_2CO_3 preparations through x=0.30 but then become erratic and finally, at x=1, superconductivity is lost.

STUDIES OF BISMUTH SUBSTITUTION [9]

After the seminal discovery of "high T_c" superconductivity in $(La,Ba)_2CuO_{4-\delta}$ above 30K [1], it was found that substitution of Sr for Ba raised the T_c to about 40K [10-12]. In apparent further attempts at substitution in this structure, $RBa_2Cu_3O_y$ ("1-2-3") compounds (where R is Y or any of a number of trivalent rare earth elements and $y \approx 7$) with a T_c of about 92K were discovered [2]. Following this, probably initially in efforts to prepare $BiBa_2Cu_3O_y$ (and with the knowledge that Bi^{3+} can replace Y and rare earth in garnets), the compounds $Bi_2Sr_2CuO_{6+x}$, $Bi_2Sr_2CaCu_2O_{8+x}$, and $Bi_2Sr_2Ca_2Cu_3O_{10+x}$ (with T_c values of about 10, 85, and 110K, respectively) were prepared [13-15]. No such compounds with Ba replacing Sr or Ca have been reported. Other superconducting compounds relevant to the current study are the copper-free Bi compounds $Ba(Pb_{1-x}Bi_x)O_3$ [16] and $(Ba_{1-x}K_x)BiO_3$ [17] with T_c values of 13K (at x=0.3) and about 30K (at x=0.4), respectively.

With various motives, there have been a number of previous studies involving $RBa_2Cu_3O_y$ and Bi^{3+}; all of these were based on preparations in air or oxygen.

When Kilcoyne and Cywinski [18] attempted to improve the intergrain contact in sintered samples of $YBa_2Cu_3O_y$ by partial substitution of Bi for Y, the Bi oxide acted as a flux, changes in the morphology of the sintered grains were observed, and the normal state resistivity was reduced by almost an order of magnitude while leaving the T_c unchanged. They asserted that Bi substitutes for Y up to at least $(Y_{0.85}Bi_{0.15})Ba_2Cu_3O_y$ (but this conclusion was apparently based only on electron microscope observations).

Jung et al. [19], on studying the system $(Y_{1-x}Bi_x)Ba_2Cu_3O_y$, found that Bi does not change the lattice parameters and concluded that there is no Bi solubility.

Spencer and Roe [20] reacted $Bi_2Sr_2CaCu_2O_{8+x}$ with $YBa_2Cu_3O_y$ and found a new nonsuperconducting (above 77K) face-centered cubic phase with a=0.855 nm which they claimed was isomorphous with the high-temperature form of $Cd_3Bi_{10}O_{18}$. Although they did not establish the exact composition of the new compound, they did state that it contained Y, Bi, and Ba. Products were superconducting with x as high as 0.8 in the formula $(YBa_2Cu_3)_{1-x}(Bi_2Sr_2CaCu_2)_xO_y$. $YBa_2Cu_3O_y$ and the new phase coexisted in the superconducting compositions. Chaffron et al.[21] studied sintering of what they called $Y_{1-x}Bi_xBa_2Cu_3O_7$ but found that a new impurity phase was formed immediately. However, the partial substitution of Bi increased the density and reduced the normal state resistivity by two orders of magnitude.

Blower and Greaves [22] tried unsuccessfully to replace Y with Bi and reported the formation of an insulating cubic perovskite-type compound Ba_2YBiO_6 with a=0.85675 nm and Bi and Y ordered on octahedral sites.

Liu et al.[23] found that Bi substituted for Cu in $YBa_2Cu_3O_y$ improves the hardness but gave no information on T_c or the x-ray diffraction pattern.

Chandrasekaran et al.[24] found that 1-10 mole % Bi_2O_3 in $YBa_2Cu_3O_y$ improved grain size and interconnectivity with no change in T_c.

Suzuki et al.[25] reported that the product of firing a composition designed to yield $YBa_2Cu_3O_y$ and 1 mole % excess Bi [i.e., $(YBa_2Cu_3O_y)_{0.99}(BiO_{1.5})_{0.01}$] had a T_c unchanged from that of $YBa_2Cu_3O_y$.

Zhuang et al.[26] assumed that Bi would substitute for Cu in $YBa_2(Cu_{1-x}Bi_x)O_y$ (with x=0-0.9). They found that Bi causes T_c to fall but that with x<0.25, zero resistance temperatures were still above 77K. For $x \geq 0.05$, a new phase, Ba_2YBiO_6, formed and had a perovskite-type structure with a=0.8549 nm. They concluded that any solubility of Bi in $YBa_2Cu_3O_y$ is small.

Yang et al.[27, 28] found that Ba_2YBiO_6 was formed when attempts were made to substitute Bi for Cu in $YBa_2Cu_3O_y$ or when $YBa_2Cu_3O_y$ reacted with Bi_2O_3. The new phase caused degradation of both superconducting and mechanical properties and the calcined powders displayed color variation ranging from black to brown as Bi content increased. When prereacted $YBa_2Cu_3O_y$ was heated with Bi_2O_3, x-ray diffraction indicated formation of Ba_2YBiO_6 and CuO even at 800°C.

Some of the above papers claim substitution of Bi for Y in $YBa_2Cu_3O_y$, but it appears likely on consideration of all of them that this was not really achieved, and that the problem is oxidation of Bi^{3+} to Bi^{5+} and resultant formation of the very stable Ba_2YBiO_6 phase when the large alkaline Ba^{2+} ion is present. We were interested in pursuing this further, however, because $(Y_{1-x}Bi_x)Ba_2Cu_3O_y$ with x from 0 to 1 would enable a study of the effect of electronegativity difference and percent covalent character on T_c. Also, from the pragmatic point of view, T_c's of the Bi-based superconductors with Ca and Sr rather than Ba have been found to be as high as 110K.

Our initial plan was to prepare tetragonal $(Y_{1-x}Bi_x)Ba_2Cu_3O_y$ (where y would presumably be in the range 6-6.5) in inert atmosphere at temperatures between about 800 and 950°C and then to insert the additional oxygen required by annealing in O_2 at temperatures from 250-500°C. It was thought that Bi^{3+} could exist and enter the $RBa_2Cu_3O_y$ lattice in the inert atmosphere and that the oxidation of this phase to orthorhombic $(Y_{1-x}Bi_x)Ba_2Cu_3O_{\sim 7}$ by oxygen at low temperatures could be carried out without permitting oxidation of Bi^{3+} and formation of the Ba_2YBiO_6.

Experimental (Bi)

Although it was planned initially to base preparations on the general formula $(Y_{1-x}Bi_x)Ba_2Cu_3O_y$, it was quickly found by x-ray diffraction that $YBa_2Cu_3O_y$ did not form in the inert gas (Ar or N_2) atmosphere desired. This observation has also been made by Parent et al.[29]. We then went to Nd in place of Y because: (1) Nd_2O_3 melts at a lower temperature than Y_2O_3 (2272°C vs. 2410°C) and should therefore be more reactive at any given temperature; (2) Noel and Parent [30] reported that Nd is more favorable than Y for the formation of $RBa_2Cu_3O_7$ because secondary phases are present in smaller quantities; and (3) the effective ionic radii [7] for 8-fold coordination of Nd^{3+} (0.1109 nm) and Bi^{3+} (0.117 nm) are much closer together than are those of Y^{3+} (0.1019 nm) and Bi^{3+}.

Compositions were at first planned to be based on $(Nd_{1-x}Bi_x)Ba_2Cu_3O_y$ with x-values from 0 to 1, but it was soon found by x-ray diffraction that an x-value of 0.5 was already far beyond the maximum possible for single-phase formation.

To prepare these materials, appropriate quantities of Nd_2O_3 (99.9%, Research Chemicals, preheated at 1200°C for one hour to decompose any $Nd(OH)_3$ present), Bi_2O_3 (99.9%, J. T. Baker Chemical Co.), $BaCO_3$ (Mallinckrodt Analytical Reagent), and CuO (Mallinckrodt Analytical Reagent) were thoroughly ground together and transferred to open high-density alumina crucibles. Synthesis was then carried out under flowing Ar or N_2 gas in a one-end-closed combustion tube inside a Harper muffle furnace with automatic temperature control ($\pm$ 10°C) at different reaction temperatures for different sample compositions. Because the melting points decreased with increasing Bi content, the various reac-tion temperatures were achieved simultaneously by placing samples at different points along the natural temperature gradient of the furnace; the gradient chosen was 880-950°C. The samples were kept at these temperatures for 10-20 hours to effect complete reaction and were then fairly rapidly cooled in the inert gas atmosphere by removing the combustion tube from the furnace with gas still flowing. Because we were expecting the tetragonal form at this point it was not necessary to cool slowly.

To prepare samples for electrical measurements, cooled preparations were then reground and shaped into 1/2-inch-diameter discs in a cylindrical die in a Carver hydraulic press at a pressure of 0.14-0.28 GPa (2.0-4.0 x 10^4 lbs/in^2). The pellets were placed in the combustion tube in the furnace at the same temperature and in the same atmosphere as in the previous reactions for 1-4 hours of sintering and were then cooled as before. To add oxygen, samples were finally annealed in air or O_2 (the latter at a flow rate of 31 cm^3/minute in the combustion tube) at 250-500°C for several hours (O_2) up to several weeks (air). Preparation of other types of samples will be described whenever required below.

Results and Discussion (Bi)

For comparison with materials to be prepared in inert atmospheres (as outlined above), $NdBa_2Cu_3O_y$ (i.e., with x=0 in $(Nd_{1-x}Bi_x)Ba_2Cu_3O_y$) was prepared by firing stoichiometric quantities of Nd_2O_3, $BaCO_3$, and CuO in air at 950°C, cooling slowly, and then heating in air or O_2 at 400°C. The product was found to have the expected $YBa_2Cu_3O_y$ structure and T_c (measured on pellets sintered in O_2 in the usual way). When some Bi was introduced, a new cubic compound formed along with the $NdBa_2Cu_3O_y$ even at x=0.1. This compound, which appeared to be analogous to the previously reported Ba_2YBiO_6, was then synthesized by firing stoichiometric quantities of reactants to yield pure Ba_2NdBiO_6, which has a brown color. Its lattice constant has been determined to be 0.8703 nm, somewhat larger than values given above for Ba_2YBiO_6 because the radius of Nd^{3+} is larger than that of Y^{3+}. As in the Y-Bi-Ba-Cu-O system [27, 28] the products in the Nd-Bi-Ba-Cu-O system displayed gradual color variation ranging from black to brown as the amount of Bi was increased. Also, as x was increased the amount of Ba_2NdBiO_6 formed increased and there was no change in the lattice constant of $NdBa_2Cu_3O_y$. This indicates that little or no Bi enters the $NdBa_2Cu_3O_y$ prepared in air and that the Bi^{3+} is oxidized to Bi^{5+}. The results of electrical measurements on pellets sintered in oxygen, shown in Figure 4 and Table II, indicate that, as Bi content increases, the normal resistivities of the samples are elevated and the conductivity changes gradually from metallic to semiconducting. Below x=0.3, the transition-temperature range becomes broader with increasing Bi content but the T_c onset temperature remains much the same. These changes are due to the presence of the nonsuperconducting Ba_2NdBiO_6. Above x=0.3, T_c decreases and the transition range is even broader. Above x=0.5, superconductivity is not observed at all.

For preparation in inert atmosphere (Ar or N_2) it was found that it was more difficult to achieve complete reaction than with the same reactants in air. Longer reaction time rather than higher temperature was adopted in order to avoid melting. X-ray diffraction patterns of materials prepared in Ar or N_2 indicate, as expected, that the $NdBa_2Cu_3O_y$ is always tetragonal rather than orthorhombic, even when cooled slowly to room temperature. An important difference between preparation in inert atmosphere rather than air is that Cu_2O is observed only in the former.

At x<0.2, it was very difficult to find Ba_2NdBiO_6 in the x-ray patterns although some impurity-phase lines were present (including those of Cu_2O) but the lattice parameters of tetragonal $NdBa_2Cu_3O_y$ remained constant. However, when these samples were heated in oxygen even as low as 300°C lines of Ba_2NdBiO_6 clearly emerged (slowly at 300°C, much more rapidly at 400°C) along with the pattern of orthorhombic $NdBa_2Cu_3O_y$ (with lattice parameters still unaffected by composition), and sample colors shifted from gray-black to dark brown. Is it possible, then, that some Bi^{3+} replaced Nd^{3+} in inert atmosphere though at too low a concentration to affect the lattice constants and that Ba_2NdBiO_6 formed on heating in air? Probably this is not the case because it does not seem that the activation energy for formation of Ba_2NdBiO_6 could be overcome at such low temperatures. Some support for this conclusion was provided by the observation of the presence of unreacted Bi_2O_3 but not of Ba_2NdBiO_6 in a mixture of Bi_2O_3 and $NdBa_2Cu_3O_y$ (with a Bi:Nd ratio of 1:5) fired

at 550°C in either air or N_2. With x≥0.3 in nominal $(Nd_{1-x}Bi_x)Ba_2Cu_3O_y$, Ba_2NdBiO_6 was observed on preparation in inert atmosphere although the x-ray patterns of this compound were weak prior to reheating in oxygen. We assume that the oxygen for any partial oxidation of Bi^{3+} to Bi^{5+} in inert atmosphere came from reduction of CuO to Cu_2O.

If Ba_2NdBiO_6 is not observed at x≤0.2 in inert atmosphere and there is no change in the lattice constant of $NdBa_2Cu_3O_y$, where is the Bi? We propose that oxygen-deficient $Ba_2NdBiO_{6-\delta}$ is present but in poorly crystallized form unless it is subsequently heated in air. This would explain why the reheating in air can be at temperatures as low as about 300°C; that is, the phase has already formed in an essentially non-crystalline form but becomes crystalline and stoichiometric on reheating in air.

To provide some confirmation that oxygen-deficient $Ba_2NdBiO_{6-\delta}$ is formed before final oxygen annealing, thermogravimetric analysis (TGA) experiments were carried out. For materials prepared in air or else in N_2 followed by annealing in air or O_2, TGA showed no mass increase as the samples were reheated in air from room temperature to 800°C in the TGA apparatus. This shows that oxygen is not absorbed by these samples, which x-ray diffraction shows to contain orthorhombic $NdBa_2Cu_3O_y$.

Samples prepared in N_2 did, however, gain mass during heating in air in the thermogravimetric analyzer (Fig. 5) and it should be noted that different samples had different onset and maximum temperatures. It is seen on comparison of oxygen-deficient $NdBa_2Cu_3O_y$ and $Ba_2NdBiO_{6-\delta}$ with nominal $(Nd_{1-x}Bi_x)Ba_2Cu_3O_y$ (with x=0.05-0.3) that the temperature of maximum oxygen absorption shifts gradually from 400 to 570°C as x increases while the temperatures of maximum oxygen absorption of $NdBa_2Cu_3O_y$ and $Ba_2NdBiO_{6-\delta}$ are at about 400 and 570°C, respectively. This indicates that defect $Ba_2NdBiO_{6-\delta}$ was already present in the Bi-containing samples before annealing in air and that, like tetragonal $NdBa_2Cu_3O_y$, it absorbs oxygen at rather low temperatures.

Figure 6 illustrates this. Nominal $(Nd_{0.8}Bi_{0.2})Ba_2Cu_3O_y$ prepared in N_2 and then annealed in O_2 at 480°C for 60 hours gains no mass on reheating in air in the TGA apparatus from room temperature to 800°C. However, a sample with the same composition prepared in

N_2 but then annealed in air at 300°C for more than a week does absorb oxygen during the TGA heating but primarily at the higher temperature required by $Ba_2NdBiO_{6-\delta}$. This indicates that the $NdBa_2Cu_3O_y$ has already acquired nearly sufficient oxygen at 300°C but that the $Ba_2NdBiO_{6-\delta}$ has not, even though it is already present. Although the $NdBa_2Cu_3O_y$ has a lower temperature of maximum oxygen absorption than does $Ba_2NdBiO_{6-\delta}$, the latter takes up some oxygen even at lower temperatures because both have nearly the same temperature of onset of oxygen absorption. However, the $Ba_2NdBiO_{6-\delta}$ requires higher temperatures to reach $\delta = 0$.

Electrical conductivity measurements (see Figures 7 and 8, and also Table III) indicate that the superconductivity of samples prepared in inert atmosphere at high temperature and then annealed in air or oxygen at lower temperatures is degraded even in the absence of Bi (i.e., at x=0 in $(Nd_{1-x}Bi_x)$-$Ba_2Cu_3O_y$). Apparently, orthorhombic $NdBa_2Cu_3O_y$ with optimum composition is not easily formed from the tetragonal phase prepared in Ar or N_2. Not only are T_c values depressed (especially at zero resistance), but also the width of the transition is increased and even the Bi-free $NdBa_2Cu_3O_y$ exhibits a negative slope for resistance vs. temperature above T_c, thereby indicating semiconducting rather than metallic behavior in the normal state. The conductivity of Bi-containing preparations is consistent with the other physical data reported above.

We therefore do not find evidence for solubility of Bi^{3+} in either orthorhombic or tetragonal $NdBa_2Cu_3O_y$. Although it is still conceivable that there is a small finite solubility (with less than 5% of the Nd^{3+} replaced), it is likely that this is not so. If such a small concentration of Bi^{3+} is indeed present in the tetragonal phase, it could perhaps be converted to the orthorhombic phase by plasma oxidation at room temperature [31], where Ba_2NdBiO_6 would presumably be unable to form. Even if this did occur, however, such a low concentration of Bi^{3+} would not be expected to have much effect on the superconducting behavior of the orthorhombic phase.

Apparently Bi and the large alkaline Ba^{2+} ion have such a strong "affinity" for each other to form the very stable Ba_2NdBiO_6 (or a poorly crystallized precursor of it) that Ba^{2+} is taken from the 1-2-3 compound. In air or oxygen, crystalline Ba_2NdBiO_6

clearly forms. Even in Ar or N_2 sufficient oxygen is made available by the reduction of Cu^{2+} to Cu^+ to oxidize at least some of the Bi^{3+} to Bi^{5+} for Ba_2NdBiO_6 formation.

LITERATURE CITED

1. J. G. Bednorz and K. A. Müller, Z. Phys. B: Condensed Matter 64, 189 (1986).

2. The first indication of such compounds was reported by M. K. Wu, J. R. Ashburn, C. J. Torng, P. H. Hor, R. L. Meng, L. Gao, Z. H. Huang, Y. Q. Wang, and C. W. Chu, Phys. Rev. Lett. 58, 908 (1987).

3. L. Suchow, J. R. Adam, and K. S. Sohn, J. Supercond. 2, 485-492 (1989).

4. M. F. Yan, W. W. Rhodes, and P. K. Gallagher, J. Appl. Phys. 63, 821-828 (1988).

5. S. X. Dou, A. J. Bourdillon, X. Y. Sun, H. K. Liu, J. P. Zhou, N. Savvides, C. C. Sorrell, K. E. Easterling, and D. Haneman; submitted paper received privately.

6. J. Wang, F. Boterel, G. Desgardin, J. M. Haussonne, and B. Raveau, Ind. Ceram. (Paris) 832, 779-785 (1988).

7. R. D. Shannon, Acta Crystallogr. A32, 751 (1976).

8. W. Wong-Ng, H. F. McMurdie, B. Paretzkin, Y. Zhang, K. L. Davis, C. R. Hubbard, A. L. Dragoo, and J. M. Stewart, Powder Diffract. 2, 191-201 (1987).

9. L. Suchow and J. Tang, J. Supercond. 4, 289-296 (1991).

10. R. J. Cava, R. B. van Dover, B. Batlogg, and E. A. Reitmann, Phys. Rev. Lett. 58, 408 (1987).

11. J.-M. Tarascon, L. H. Greene, W. R. McKinnon, G. W. Hull, and T. H. Geballe, Science (Washington, D.C.) 235, 1373 (1987).

12. D. W. Capone, D. G. Hinks, J. D. Jorgensen, and K. Zhang, Appl. Phys. Lett. 50, 543 (1987).

13. C. Michel, M. Hervieu, M. M. Borel, A Grandin, F. Deslandes, J. Provost, and B Raveau, Z. Phys. B: Condensed Matter 68, 421 423 (1987).

14. H. Maeda, Y. Tanaka, M. Fukutumi, and T Asano, Jpn. J. Appl. Phys. 27, L209 (1988).

15. J.-M. Tarascon, W. R. McKinnon, P. Barboux, D. W. Hwang, B. G. Bagley, L. H. Greene, G. W. Hull, Y. LePage, N. Stoffel, and M. Giroud, Phys. Rev. B38, 8885-8892 (1988).

16. A. W. Sleight, J. L. Gillson, and P. E. Bierstedt, Solid State Commun. 17, 27 (1975).

17. R. J. Cava, B. Batlogg, J. J. Krajewski, R. Farrow, L. W. Rupp, Jr., A. E. White, K. Short, W. F. Peck, and T. Kometani, Nature (London) 332 (6167), 814-815 (1988).

18. S. H. Kilcoyne and R. Cywinski, J. Phys. D.: Appl. Phys. 20, 1327-1329 (1987).

19. J. Jung, J. P. Franck, W. A. Miner, and M. A.-K Mohamed, Phys. Rev. B37, 7510-7515 (1988).

20. N. D. Spencer and A. L. Roe, Chap. 12 of "Chemistry of High-Temperature Superconductors II", ACS Symposium Series 377, ed. by D. L. Nelson and T. F. George; American Chemical Society, Washington, D.C. (1988).

21. L. Chaffron, J. P. Mercurio, and B. Frit, Ind. Ceram. (Paris) 835, 122-124 (1989).

22. S. K. Blower and C. Greaves, Solid State Commun. 68, 765-767 (1988).

23. Y. Liu, J. Li, H. Liu, and W. Su; Jilin Daxue Ziran Kexue Xuebo 1989 (1), 75-77.

24. K. D. Chandrasekaran, U. V. Varadaraju, A. Baradarajan, and G. V. Subba Rao, Bull. Mater. Sci. 12, 81-93 (1989).

25. T. Suzuki, T. Yamazaki, A. Koukitu, M. Maeda, H. Seki, and K. Takahashi, J. Mater. Sci. Letters 7, 926-927 (1988).

26. J. Zhuang, W. Su, H. Liu, Y. Wang, and J. Zhou, J. Less-Common Metals 149, 427-434 (1989).

27. N. Yang, R. S. Liu, W. N. Wang, Y. H. Chao, H. C. Lin, and P. T. Wu, Physica C162-164, 71-72 (1989).

28. N. Yang, R. S. Liu, W. N. Wang, and P. T. Su, J. Mater. Sci. 25, 4758-4762 (1990).

29. L. Parent, B. Champagne, K. Cole, and C. Moreau, Supercond. Sci. Technol. 2, 103-106 (1989).

30. D. Noel and L. Parent, Thermochim. Acta 147, 109-117 (1989).

31. B. G. Bagley, L. H. Greene, J.-M Tarascon, and G. W. Hull, Appl. Phys. Lett. 51, 622-624 (1987).

Table I

T_c Values for "Set 1" (from Li_2CO_3, 910-925°C)

	x in $YBa_2Cu_{3-x}Li_xO_y$	Percent Cu replaced by Li	Temperatures (K)		
			Onset	Midpoint*	Zero Resistance
	0	0	97	91	88.5
	0.01	0.33	95	92	91
	0.02	0.67	100	93	89
	0.03	1	94	91	90
	0.06	2	102	92	90
	0.09	3	97	93	91.5
	0.15	5	93	90	85
**	0.30	10	91	83	72.5
***	0.50	17	78	67.5	46
	1.0	33	80	58	30

   * Temperature at resistance midway between zero and the resistance just prior to onset (that is, the temperature at half the resistance just prior to onset).
  ** Resistance falls on cooling to about 155K, then rises before the major drop at T_c.
*** Resistance falls on cooling to about 200K, then rises before the major drop at T_c.

Table II

T_c Values of Nominal $Nd_{1-x}Bi_xBa_2Cu_3O_y$ Prepared in Air and then Annealed in O_2 at 400°C for 40 hours (Key data from Figure 4)

x	Temperatures (K)		
	Onset	Midpoint*	Zero Resistance
0	92	87	81
0.1	92	89	78
0.2	93	85	77
0.3	91	86	72

*Temperature at resistance midway between zero and the resistance just prior to onset (that is, the temperature at half the resistance just prior to onset).

Table III

T_c Values of Nominal $Nd_{1-x}Bi_xBa_2Cu_3O_y$ Prepared in N_2 and then
Annealed in Air or O_2 (Key data from Figures 7 and 8)

Fig. No.	x	Temperatures (K)		
		Onset	Midpoint*	Zero Resistance
7	0	92	78	57
7	0.05	94	78	46
7	0.1	88	37	-
7	0.2	85	26	-
8	0.1(Curve 1)	70	51	26
8	0.1(Curve 2)	30	~0	-

* Temperature at resistance midway between zero and the resistance just prior to onset (that is, the temperature at half the resistance just prior to onset). The last value in this column was calculated by extrapolation.

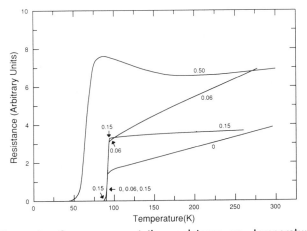

Figure 1. Some representative resistance vs. temperature data. Numbers at curves are values of x in $YBa_2Cu_{3-x}Li_xO_y$.

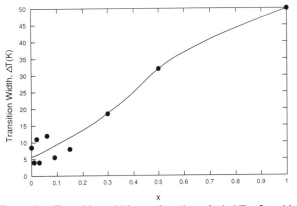

Figure 3. Transition width as a function of x in $YBa_2Cu_{3-x}Li_xO_y$.

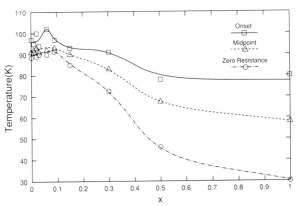

Figure 2. Onset, midpoint, and zero resistance temperatures as a function of x in $YBa_2Cu_{3-x}Li_xO_y$.

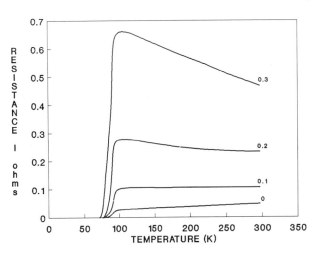

Figure 4. Electrical resistance vs. temperature for nominal $(Nd_{1-x}Bi_x)Ba_2Cu_3O_y$ prepared in air and then annealed in O_2 at 400°C for 40 hours. Numbers at curves are x-values.

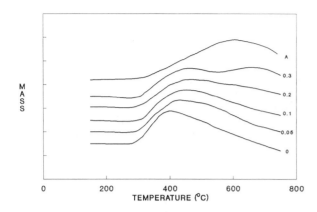

Figure 5.   Thermogravimetric analysis in air of samples after preparation in N_2. Curve A: $Ba_2NdBiO_{6-\delta}$. All others are actual or nominal $(Nd_{1-x}Bi_x)Ba_2Cu_3O_y$ compositions with x-values shown. (Mass values on y-axis are relative; curves are arranged for comparison).

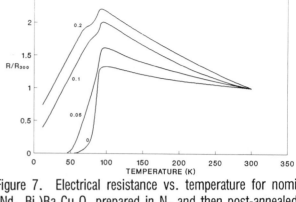

Figure 7.   Electrical resistance vs. temperature for nominal $(Nd_{1-x}Bi_x)Ba_2Cu_3O_y$ prepared in N_2 and then post-annealed in O_2 at 480°C for 40 hours. Numbers at curves are x-values. R/R_{300} is resistance divided by resistance of the same sample at 300K. (Presented in this fashion because large increases in resistance levels with increasing Bi content do not permit all the curves to be displayed directly on one linear scale).

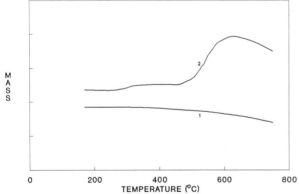

Figure 6.   Thermogravimetric analysis in air of nominal $(Nd_{0.8}Bi_{0.2})Ba_2Cu_3O_y$ samples after preparation in N_2 and post-annealing. Curve 1: Post-annealed in O_2 at 480°C for 60 hours. Curve 2: Post-annealed in air at 300°C for more than a week. (Mass values on y-axis are relative; curves are arranged for comparison).

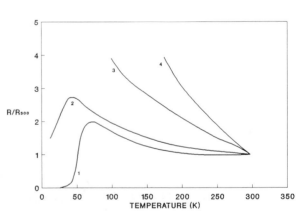

Figure 8.   Electrical resistance vs. temperature for nominal $(Nd_{1-x}Bi_x)Ba_2Cu_3O_y$ and then post-annealed.
Curve 1: x = 0.1, post-annealed at 340°C in air for 20 hours and then in O_2 for 20 hours.
Curve 2: x = 0.1, post-annealed in air at 340°C for 40 hours.
Curve 3: x = 0.05, post-annealed in air at 300°C for more than a week.
Curve 4: x = 0.2, post-annealed in air at 300°C for more than a week.
R/R_{300} is resistance divided by resistance of the same sample at 300K. (Presented in this fashion because large differences in resistance levels do not permit all the curves to be displayed directly on one linear scale).

THE NATURE OF OXYGEN IN OXIDE SUPERCONDUCTORS

N. A. Gokcen ■ Albany Research Center, U.S. Bureau of Mines, U.S. Department of the Interior, 1450 Queen Avenue SW, Albany, Oregon 97321, U.S.A.

The dissolution process for gaseous oxygen in $YBa_2Cu_3O_{6+2r}$ and in La_2CuO_{4+d}, where r and d are the numbers of occupied sites, is $0.5\ O_2(g) =$ [dissolved O]. The statistical thermodynamic equation for the dissolution of oxygen in $YBa_2Cu_3O_{6+2r}$ is

$$\Delta G^\circ = -RT\ln[r/\sqrt{[P(O_2)]}]$$
$$= -88,300 + 56.55T \qquad (1)$$
$$- (75,200 - 218.3T)r - RT(1 - r),$$

where ΔG° is the standard Gibbs energy change in J/mol[0], $R = 8.3144$ J/mol·K, T is temperature in K, and $P(O_2)$ is oxygen pressure in atm. The corresponding equation for La_2CuO_{4+d} is

$$\Delta G^\circ = -RT\ln[d/\sqrt{[P(O_2)]}]$$
$$= -36,500 + 92T. \qquad (2)$$

Analyses of the published data suggest that for high-temperature superconductivity (i) a layered-type crystal structure appears to be essential, (ii) excess oxygen plays a major role, and (iii) at least one element capable of forming a peroxide, i.e., -O-O- chains, is also essential. The copper atoms play an important role in all the new oxide superconductors; how-ever, other elements with their oxides having odd numbers of d-electrons might well lead to new higher T_C superconductors at high oxygen pressures. This research is part of the effort for advancement of technology at the U.S. Bureau of Mines.

The physicochemical nature of oxygen in the recently discovered multicomponent oxide superconductors is not clearly understood [1-11], largely due to lack of understanding the nature of oxygen in the single-metal oxides (Gokcen et al. in Reference 11). Some oxides such as MgO, CuO, Al_2O_3, Y_2O_3, and La_2O_3 retain very nearly perfect stoichiometries up to a few hundred bars of oxygen pressure $P(O_2)$ at temperatures T up to about 1,300 K, but alkali and alkaline earth metals may form oxides of varying stoichiometries as in K_2O, K_2O_2, K_2O_4, Ba_2O, BaO, and BaO_2. In peroxides such as K_2O_2 and BaO_2, the maximum valences of 1 for K and 2 for Ba require that -O-O-, or O_2^{2-} entities exist in these compounds. Such entities are correctly interpreted as the oxygen 2p-band holes in peroxides and their compounds. It seems likely that as $P(O_2)$ increases up to the neighborhood of 30 kilobars, most stoichiometrically stable oxides such as CuO, Y_2O_3, and La_2O_3 may dissolve significant amounts of oxygen above 1,300 K. It appears that such dissolved oxygen species are also of the O_2^{2-}-type, as will be discussed later.

The multioxide compounds (or phases) exhibit remarkable changes in their physicochemical and electrical properties with changes in their oxygen contents as affected by the oxygen pressures at various temperatures. Thus $YBa_2Cu_3O_{6+2r}$ and La_2CuO_{4+d} are superconductors when $6 + 2r = 7$ and $4 + d = 4.10$, but both compounds are nonsuperconductors when r and d are zero. The effects of oxygen on the more recently discovered Bi- and Tl-based superconductors are much less known at present. However, it is the author's opinion that O_2^{2-} also plays significant roles in these superconductors.

It was shown experimentally and presented in a recent paper that r in $YBa_2Cu_3O_{6+2r}$ and d in BaO_{1+d} vary with $P(O_2)$ of comparable magnitudes at approximately the same temperatures (Gokcen et al. in Reference 11). Further, it was determined experimentally that the oxygen stoichiometry of the remaining component oxides, Y_2O_3 and CuO, did not change with $P(O_2)$ up to about 150 bars and about 1,000 K. In addition, a well-characterized cuprate, $Y_2Cu_2O_5$, also did not change in oxygen composition from 1 to 150 bars at about 1,070 K. It was therefore concluded that Y and Cu in $YBa_2Cu_3O_{6+2r}$ (hereafter Y123 when no emphasis is needed for r) remained trivalent and divalent, respectively, but BaO in the compound formed a peroxide BaO_{1+d}, in which some of the oxygen atoms formed -O-O- bonds since the valence of Ba cannot exceed 2. The presence of -O-O- or O_2^{2-} in Y123 is supported by a number of investigators who have shown that trivalent

copper, Cu^{3+}, does not exist. Thus, X-ray photoelectron spectroscopy and ultraviolet photoemission spectroscopy support the presence of only Cu^{2+} [12-14] (see also Bianconi in Reference 7). Further, Curtiss and Shastry [10] concluded that the removal of the third electron from Cu to get Cu^{3+} is energetically difficult. For a theoretical argument against trivalent copper, see Klein et al. [8]. Despite these results and the correct interpretations for the divalency of Cu, numerous investigators have incorrectly postulated the existence of trivalent copper in Y123 and in other cuprates (see, for example, a number of papers in References 1-11). This postulate on the trivalent copper is based on the crystallographic location of copper in the perovskite-type oxide compounds [1-11].

DISSOLUTION OF OXYGEN IN $YBa_2Cu_3O_{6+2r}$

The oxygen atoms in $YBa_2Cu_3O_6$ are shown in Figure 1(A), where the structure is tetragonal. Likewise, the orthorhombic structure for $YBa_2Cu_3O_7$ is shown in Figure 1(B). For simplicity, these structures are shown without the interplanar distortion. The interstitial oxygen atoms along the b-axis in $YBa_2Cu_3O_7$ form a chain of Cu-O-Cu-O, and these chains are interpreted to be responsible for superconductivity. The interstitial sites along the a-axis, and possibly along the remaining sites between the metal atoms, could be occupied by oxygen at sufficiently high oxygen pressures. Complete occupancy of the interstitial sites along the a- and b-axis would yield $YBa_2Cu_3O_8$. However, a compound having more than 7.3 atoms of oxygen has not yet been reported with full characterization of its properties. Attempts to obtain nearly 8 atoms of oxygen might possibly lead to the dissociation and unusual phase transformations of the resulting compound.

The equilibrium oxygen pressures at various temperatures are correlated in Figure 2 (see Gokcen et al. in Reference 11). The analytical uncertainty in 2r is likely to be about ±0.03 according to Nazzal et al. [7], although the variation in 2r can be determined with an accuracy of better than ±0.01 on a good differential thermogravimetric analyzer. Further, at the values of T less than 700 K and $P(O_2)$ less than 0.01 atm, experimental difficulties arise in equilibration. Therefore, it is possible to formulate the results in Figure 2, on the basis of these limitations, by using the interstitial dissolution equations [15]. The dissolution of diatomic gases in metals and oxides obey the Sieverts law and its extended version based on statistical thermodynamic theories [15-16]. This law and its extension require that $O_2(g)$ dissolves in Y123 as follows:

$$0.5O_2(g) = [O]. \qquad (3)$$

The upper and lower basal planes can accommodate two atoms per $YBa_2Cu_3O_6$, hence we set c in Reference 15 to 2; thus, the number of vacant sites per crystal in Figure 1(B) is 2, and y = r in Equation 6.1 and 6.20 of Reference 15. The corresponding standard Gibbs energy change, ΔG^o, for Reaction 1 is

$$\Delta G^o(J/g.atom\ O) = -RT\ln K_p$$
$$\equiv RT\ln \frac{P(O_2)^{0.5}}{r} = -88,300 + 56.55T$$
$$- (75,200 - 218.3T)r - RT\ln(1-r) \qquad (4)$$

where $K_p = r/\sqrt{P(O_2)}$ for Reaction 1 with $P(O_2)$ in atm. The numerical terms in this equation were obtained in the present paper by an optimum fit to the data represented in Figure 2 within a probable error of ±0.03 in 2r. Considerably improved data are necessary to obtain a relationship more reliable than Equation 4. Such

data are currently being obtained at the Bureau of Mines.

Equation 4 yields $P(O_2) = 0.166$ atm for r = 0.4 and $P(O_2) \approx 330$ atm for r = 0.6 at 800 K. Thus, with increasing values of r, $P(O_2)$ increases quite rapidly. Higher pressure data than those in Figure 2 are necessary to calculate more reliable values of r from $P(O_2)$ and vice versa for a given temperature.

The effects of oxygen on the critical superconducting temperatures (T_c) of Y123-type compounds have been determined by numerous investigators [7-11]. Synthesis, heat treatment, and contamination affect the values of T_c for the same superconductor with the same stoichiometry. Therefore, the results of various investigators are not always in satisfactory agreement. A recent set of carefully obtained data by Wong-Ng et al. [11] on five such compounds is reproduced in Figure 3, showing the variation of T_c with the oxygen composition. At least one plateau having a lower value of T_c exists for all the compounds, and another plateau at a higher value of T_c exists for Y123 and Er123. It seems possible that much higher values of T_c are achievable for Gd123 and Sm123 with a slight increase of oxygen compostion beyond 7, because the curves are steep and the second plateau might not exist up to 7.2 in oxygen contents. For this reason, an investigation of T_c and r as affected by $P(O_2)$ is being carried out by the author and his associates.

Neutron diffraction results show that the oxygen atoms randomly occupy the sites along the b-axis for 2r ≤ 1 and the electron diffraction data show that short-range order exists as determined by Wong-Ng et al.[11]. The locations of oxygen atoms when 2r is larger than 1 is controversial. However, the sites along the a-axis appear to be the possible locations for these excess oxygen atoms [5].

The orthorhombic structure is not essential for superconductivity in Y123-type compounds. Thus Wong-Ng et al. [11] have shown that a tetragonal form of Y123 is superconducting, whereas an orthorhombic form of $ErBa_2Cu_3O_{7+2r}$ is nonsuperconducting. Similar conclusions are also true for doped La_2CuO_4-type superconductors, as will be discussed later.

DISSOLUTION OF OXYGEN IN La_2CuO_{4+d}

Stoichiometric La_2CuO_4 with d = 0 can be prepared under 1 bar of $O_2(g)$ pressure (1 bar ≈ 1 atm in this section). It has the tetragonal K_2NiF_4 structure above 530 K and orthorhombic below this temperature [17-18]. The tetragonal form is shown in Figure 4(A). Pure $La_2CuO_{4.00}$ is a nonsuperconductor down to 0 K. Excess oxygen can be dissolved in this compound at high oxygen pressures and at about 600° C. Oxygen enters in the crystal as shown in Figure 4(B) and forms a superconducting phase below 37.5 K. The crystal structure of a sample having d = 0.03 (T_C = 37.5 K) was determined by neutron powder diffraction [19]. The results showed that below 430 K, the tetragonal I4/mmm phase transforms into a single orthorhombic Bmab phase, and below 320 K, this phase transforms into two orthorhombic phases, a nonsuperconducting Bmab phase ($La_2CuO_{4.00}$), and a superconducting Fmmm phase richer in oxygen. However, subsequent single crystal neutron diffraction data for $La_2CuO_{4.032}$ showed an orthorhombic Cmca phase at 15 to 300 K, without detecting a second phase [20].

The excess oxygen atoms form -O-O-, corresponding to the formation of holes in the oxygen 2p-band as correctly interpreted by Chaillout et al. [20]. The chemical association of

these oxygen atoms is probably with La since La is a neighbor of Ba in the periodic chart. This association could possibly be ascertained by first studying the stoichiometries of pure La_2O_3 and CuO under very high oxygen pressures. Such data are not yet available.

The compound La_2CuO_{4+d} is not a superconductor for zero and negative values of d, i.e., d ≤ 0. For d < 0, or for oxygen deficiency, the more easily reducible component is CuO instead of La_2O_3, and CuO would then form two species, Cu_2O and CuO, as the thermodynamic components of La_2CuO_{4+d} (Gokcen et al., Reference 11).

The dissolution of oxygen in carefully prepared samples and the effects of oxygen were investigated recently in Sandia Laboratories [21-23] and then jointly with Argonne, Los Alamos, Bell, and European Laboratories [19-20]. The latter publications, along with the earlier publications (References 24-26, and Demazeau et al. in 7) contain interesting results. The samples in these investigations were oxygenated at up to 3 kilobars of oxygen at 873 K for 12 to 48 h, followed by slow cooling to 300 K at 100 K/h. The results are therefore not necessarily equilibrium solubilities for a definite temperature. However, the diffusion of oxygen in and out of La_2CuO_4 is not rapid, therefore, it is possible to treat the resulting data approximately as the equilibrium oxygen contents in accord with Reaction 1. The value of d for samples prepared by various researchers in air and in 1 bar of oxygen disagree slightly as summarized by Jorgensen et al. [19], i.e., d = 0.01 and d = 0.01 to 0.05. Similarly prepared samples by Demazeau et al. [7]) contained d = -0.04, after correction for a small copper deficiency. A recent careful investigation [27] showed that furnace-cooled samples in air contained d = 0.00 ±0.01, in accord with the rough average of the preceding values. As-prepared samples, heated to 500° C in vacuum or in flowing inert gas for 30 min were assumed to be exactly $La_2CuO_{4.00}$ [19-21]).

In a joint effort by Sandia and Argonne Laboratories [23] the following two sets of data were obtained: Set (1): 100 bars of oxygen saturation at 580° C, followed by slow cooling yielded d = 0.03; and Set (2): 3,000 bars of oxygen saturation at 600° C, followed by slow cooling yielded d = 0.13 ± 0.02. Further, as Set (3), Zhou, Sinha, and Goodenough [27] measured the equilibrium solubility of $O_2(g)$ at 23 kbars by retaining their samples at 1,073 K for 1 h, and quenching them to obtain d = 0.143. The relative errors in d = 0.13 of Set (2) and d = 0.143 of Set (3) are smaller than in d = 0.03 of Set (1) for a possible uncertainty of Δd = ±0.02. Therefore, we use the data of Set (2) and Set (3) to write the following equilibrium constants, K_p:

$$K_p = \frac{d}{\sqrt{P(O_2)}} = \frac{0.13}{\sqrt{3,000}} = 0.0024, \quad (5)$$

$$\text{and } K_p = 0.00094.$$

The corresponding approximate standard Gibbs energy, $\Delta G°$, is

$$\Delta G° = -RT\ln K_p \approx -36,500 + 92T. \quad (6)$$

The value of d for $P(O_2)$ = 100 bars and 580° C, calculated from $\Delta G°$, is 0.027, in close agreement with 0.03 ±0.01 in Set (1). Similarly calculated value of d for 1 bar and 600° C is 0.0024, which is more reliable than the experimental value, d = 0.00 ±0.01. Equation 6 is the first correlation of d and $P(O_2)$ in La_2CuO_{4+d} as a function of temperature.

OXYGEN IN $La_{2-x}Sr_xCuO_{4+d}$

The number of oxygen atoms for perfect stoichiometry in Sr-doped lanthanum cuprate is 4 - 0.5 x, and the number of excess oxygen atoms is (4 + d) - (4 - 0.5 x) = 0.5 x + d. Each excess oxygen atom forms one O_2^{2-} or two holes; hence, the number of holes, p, is twice the excess oxygen atoms, i.e.,

$$p = x + 2d. \quad (7)$$

Torrance et al. [17-18, 28-30] prepared samples of LaSrCuO containing up to x = 0.20 in 1 bar of oxygen by slow cooling, and x = 0.24 to 0.40 by annealing at 873 K under 100 bars followed by slow cooling. High oxygen pressures, $P(O_2)$, were necessary to increase the number of holes, p, with increasing x. The resulting data are summarized in Table 1 with the column for the orthorhombic to tetragonal transition temperatures from Torrance et al. [18]. For 0.19 ≤ x ≤ 0.24, the compound is tetragonal and superconducting, whereas for 0 ≤ x ≤ 0.19, it is orthorhombic and superconducting, provided that p > 0. For d = 0.03 - 0.13 with x = 0, p = 0.06 to 0.26, and the value of T_C = 37.5 K, referred to earlier for undoped LaCuO, is in approximate agreement with Table 1. For x = 0 and d = 0.032, the volume fraction of superconductivity was determined to be 70% [20] while for x = 0 and d = 0.13, it was about 30% [21]. Indications are that the history of each sample and other factors might play important roles in the volume fraction of superconducting material in the LaCuO as well as LaSrCuO. These problems are yet to be settled [31]. Further, the values of p need to be measured at a few kilobars for LaSrCuO compounds to ascertain the effects of higher oxygen contents.

TABLE 1. — Hole concentration p and superconducting critical temperature T_C at 1 and 100 bars of oxygen for $La_{2-x}Sr_xCuO_{4+d}$ (Sources: Huang et al. [28] and Torrance et al. [17-18]

| Hole concentration, p | | | T_C, K | Orth→Tet† |
x	1 bar	100 bars		T, K
0.00	0.09*	--	20	530
0.04	0.12	--	27	--
0.08	0.15	--	35	340
0.12	0.18	--	36	--
0.16	0.22	--	38	160
0.19	--	--	36	36††
0.20	0.23	--	37	~0†
0.24	0.27	0.29	28(18)**	--
0.28	0.31	0.32	<5	--
0.32	0.31	0.38	<5	--
0.36	0.30	0.40	<5	--
0.40	0.32	0.40	<5	--

* The value for Equation 3 is 0.0048 for 1 bar and 0.048 for 100 bars, since all the samples were prepared at 600° C.

**18 K in parentheses is for 100 bars, suceeding values, <5 are for both pressures.

† Orthorhombic to tetragonal transition temperatures on heating were scaled from the smoothed curve of Figure 4 in Torrance et al. [18].

††For $0 \leq x \leq 0.19$ below each temperature the orthorhombic form is stable. For x > 0.19, the structure is tetragonal at all temperatures. For p in excess of 0.32, $T_C = 0$, and normal metallic behavior is observed [30].

Table 1 shows that the concentration of holes increases with increasing x for the same pressure. This is understandable from the viewpoint that the increasing concentration of Sr permits increasing values of p because Sr itself can readily form additional p by forming its peroxide, SrO_2. The data in Table 1 and further analysis in Figure 1 of Torrance et al. [30] suggest the following approximate relationship for 600° C:

$$p \approx 0.0048 \, [P(O_2)]^{0.5} + x;$$
$$[P(O_2) \leq 100 \text{ bars}, x < 0.4]. \qquad (8)$$

It is evident that beyond $x \approx 0.28$ for Sr and $p \approx 0.31$, the superconductivity disappears for 100 bars despite the high values of p. It would be interesting to obtain additional data at much higher pressures for $x \geq 0.28$ to assess further the effects of Sr concentration.

OXYGEN IN OTHER SUPERCONDUCTORS

The more recently discovered high-temperature superconductors [31-33] can be summarized by the general formula $A_2B_2Ca_{n-1}Cu_nO_{2n+4+d}$, usually with A/B = Bi/Sr or Tl/Ba, and n = 1, 2, or 3. (For an excellent summary, see Torardi et al. and Tarascon et al. in Reference 10.) There are other related compounds that are not discussed in the present paper.

The Bi/Sr compounds are orthorhombic for n = 1 and 2 (Torardi et al. in 10), or pseudotetragonal for n = 1, 2, and 3 (Tarascon et al. in 10). For n = 3 (or Bi2223), $T_C \approx 110$ was observed. The chemical nature of oxygen in the Bi compounds is not clearly known because Bi is capable of forming Bi_2O_5 at high pressures. At ordinary pressures, Bi_2O_5 is not stable [34], and no thermodynamic data exist for this compound. Preliminary data on the properties of these compounds indicate the disappearance of superconductivity in samples processed at 900 K and 150 bars [35]. Substantial increases in p for Bi2212 first increase T_C and then decrease it, as summarized by Torrance et al. [17].

The compounds containing Tl are difficult to investigate due to the volatility and toxicity of Tl and its oxides. For n = 3 (Tl2223), the compound is tetragonal and attains T_C = 125 K. Tl2223 has close to p = 0.15

[17], and it is possible to inject oxygen to increase T_C by probably 20 K.

CONCLUDING REMARKS

High-temperature superconductivity is encountered in about six different structures, all related to the low dimensional perovskites or layered structures (Raveau *et al.* in References 7 and 10). Not all the layered structures (resembling other superconducting layered structures) are superconducting, e.g., $La_{2x}Sr_{1+x}Cu_2O_{6+d}$ is not superconducting. Substitution of Group IIIA elements for Y in Y123 does not substantially change the highest attainable value of T_C; partial substitution for Cu may or may not lower T_C, and partial substitution of halogens for oxygen may have little effect on T_C [1-11]. However, increasing the oxygen contents of a superconductor, generally increases T_C substantially. The optimal increase depends on the optimal effects of hole concentration, p, on the compound. Complete validity of this conclusion requires further data. The picture is further complicated in that some n-type or oxygen-deficient oxide compounds are superconducting. However, T_C of n-type superconductors discovered thus far is considerably lower than T_C of p-type superconductors [36].

The cuprate-type superconductors received considerable recent attention, leading to the discovery of T_C approaching approximately 130 K. In the cuprates, the outer electronic structure of Cu is d^9, possibly signifying that an odd number of d electrons is necessary if other new superconducting compounds are to be discovered. The oxide forms of $Mn(d^3)$, $Fe(d^5)$, $Ni(d^7)$, $Ru(d^5)$ might possibly be the candidates if still higher T_C were to be explored. In the Sixth Period of the elements, the f electrons come into play, and no such statements can be made regarding the usefulness of these elements as substitutes for Cu. Further, atomic size considerations, alkali and alkaline earth elements as the peroxide-forming components, and high oxygen pressures need to be investigated if new higher T_C superconductors are to be discovered.

LITERATURE CITED

1. Gubser, D.V. and M. Schluter, eds., *High Temperature Superconductors*, Mater. Res. Soc. (1987).

2. Smothers, W.J., ed., *Ceramic Superconductors*, American Ceramic Society **2** (3B) (1987).

3. Nelson, D.L., M.S. Wittingham, and T.F. George, eds., *Chemistry of High-Temperature Superconductors*, American Chemical Society Symposium Series 351 (1987).

4. Nakajima, S. and H. Fukuyama, *Jpn. J. Appl. Phys. Series 1* (1988).

5. Poole, Jr., C.P., T. Datta, and H.A. Farach, *Copper Oxide Superconductors*, Wiley-Interscience (1988).

6. Yan, M.F., ed., *Ceramic Superconductors II*, American Ceramic Society (1988).

7. Muller, J. and J.L. Olsen, eds., *Superconductivity*, North-Holland (1988).

8. Nelson, D.L. and T.F. George, eds., *Chemistry of High-Temperature Superconductors II*, American Chemical Society Symposium Series 377 (1988).

9. Simon, R. and A. Smith, *Superconductors*, Plenum Press (1988).

10. Metzger, R.M., ed., *High Temperature Superconductivity*, Gordon and Breach (1989).

11. Whang, S.H. and A. DasGupta, eds., *High Temperature Superconducting Compounds: Processing and Related Properties*, TMMS-AIME Symposium (1989).

12. Steiner, P. *et al.*, *Z. Phys. B, Condensed Matter*, **67**, 19 (1987).

13. Sarma, D.D. *et al.*, *Phys. Rev. B.*, **36**(4), 2371 (1987).

14. Chakraverty, B.K., D.D. Sarma, and C.N.R. Rao, *Physica C,* **156**, 413 (1988).

15. Gokcen, N.A. *Statistical Thermodynamics of Alloys*, Plenum Press, chap. 6 (1986).

16. Gokcen, N.A. *Thermodynamics*, Techscience Inc., 331, (1975).

17. Torrance, J.B. *et al.*, Proceedings, Int. HTSC Conf. Stanford, *Physica C* (in press, 1990).

18. Torrance, J.B. *et al.*, *Phys. Rev. B*, **40**(13), 8872 (1989).

19. Jorgensen, J.D. *et al.*, *Phys. Rev. B*, **38**(16), 11337 (1988).

20. Chaillout, C. *et al.*, *Physica C*, **158**, 183 (1989).

21. Schirber, J.E. *et al.*, *Physica C*, **152**, 121 (1988).

22. Rogers, Jr., J.W. *et al.*, *Phys. Rev. B*, **38**(7), 5021 (1988).

23. Shinn, N.D. *et al.*, to be published in *Am. Inst. Phys. Proceedings No. 182* (1989).

24. Beille, J. *et al.*, *C. R. Acad. Sci.* **304**, Serie II(18), 1097 (1987).

25. Grant, P.M., *Phys. Rev. Lett.*, **58**, 2482 (1987).

26. Sckizawa, K. *et al.*, *Jpn. J. Appl. Phys.* **26**, L840 (1987).

27. Zhou, J., S. Sinha, and J.B. Goodenough, *Phys. Rev. B, Condensed Matter*, **39**, 12331 (1989).

28. Huang, T.C. *et al.*, *Powder Diffr.*, **4**(3), 152 (1989).

29. Tokura, Y. *et al.*, *Phys. Rev. B.*, **38**(10), 7156 (1988).

30. Torrance, Y.B. *et al.*, *Phys. Rev. Lett.* **61**(9), 1127 (1988).

31. Schirber, J.E., private communication with author, Sandia National Laboratories (Oct. 1989).

32. Maeda, H., Y. Tanaka, M. Fukutomi, and T. Asano, *Jpn. J. Appl. Phys.* **27**, L209 (1988).

33. Chu, C.W., J. Bechtold, L. Gao, P.H. Hor, and Y.Y. Xue, *Phys. Rev. Lett.* **60**, 941 (1988).

34. Sheng, Z.Z. and A.M. Hermann, *Nature*, **332**, 138 (1988).

35. Cotton, F.A. and G. Wilkinson, *Advanced Inorganic Chemistry*, Wiley-Interscience, 456 (1980).

36. Hsu, C.-H. and N.A. Gokcen (to be published).

37. Tokura, Y., H. Tagaki, and S. Uchida, *Nature*, **337**, 345 (1989).

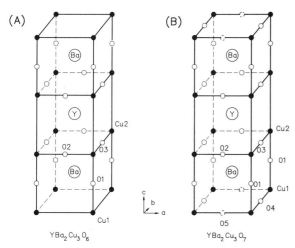

Figure 1. Crystal structures of YBa₂Cu₃O₆ and YBa₂Cu₃O₇. Small open circles are O, and small full circles are Cu. O4 sites are fully occupied in (B). Oxygen atoms in excess of 7 occupy O5 sites according to Poole, Datta, and Farach [5].

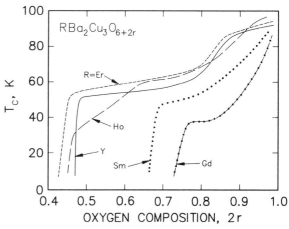

Figure 3. Variation of superconducting critical temperature, T_c, with oxygen contents for RBa₂Cu₃O₆₊₂ᵣ, where R = Er, Y, Ho, Sm, and Gd. (Adapted with permission from Wong-Ng et al. [11].)

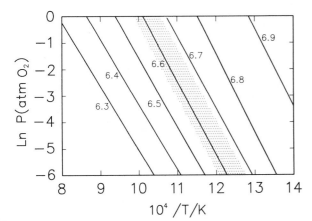

Figure 2. Variation of oxygen contents with pressure and temperature for YBa₂Cu₃O₆₊₂ᵣ. The number on each line represents 6 + 2r. The shaded strip indicates the approximate location of tetragonal to orthorhombic transition from left to right. Gokcen et al. [11].

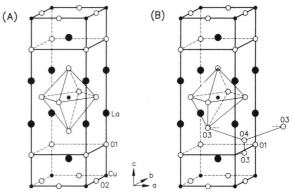

Figure 4. Crystal structures of (A) tetragonal La₄Cu₂O₈.₀₀, and (B) La₄Cu₂O₈₊₂d. Small open circles are O, and small full circles are Cu. The position of excess oxygen O4 is shown in (B) as a pseudotetragonal cell. From Chaillout et al. [20].

FORMATION OF Y-Ba-Cu (1-2-3) HYDROUS OXIDE PRECURSOR POWDERS IN THE ELECTRIC DISPERSON REACTOR†

Michael T. Harris, Timothy C. Scott, Osman A. Basaran, and Charles H. Byers ■ Chemical Technology Division, Oak Ridge National Laboratory††, Oak Ridge, Tennessee 37831-6224

The Electric Dispersion Reactor (EDR) uses a physicochemical technique to produce composite oxide ceramic precursor materials. Ultrafine (0.1- to 2-μm) spherical particles are synthesized in the EDR from aqueous/organic emulsions of solutions containing a mixture of cupric chloride, yttrium nitrate, and barium nitrate (approximately 3:1:2 molar ratio) and a continuous organic phase containing ammonia. The sintered powder forms the green phase, which indicates a deficiency in barium. This deficiency was due to the solubilization of barium hydroxide both in the continuous phase and during the wash step.

INTRODUCTION

The synthesis of ultrafine ceramic precursor powders by physicochemical techniques has gained increasing attention in recent years (1). In comparison with the strictly chemical method of homogeneous precipitation, these methods which include spray pyrolysis (2, 3), aerosol synthesis (4), electrostatic atomization of liquid drops into air (5), and precipitation in emulsions (6), offer the advantage of physically controlling the particle size by creating localized reaction zones (i.e., microreactors) in which precipitation/gelation occurs. These methods are especially attractive in the synthesis of mixed hydrous metal-oxide precursor powders when intimate mixing is required on the submicron level.

Recently, a new concept, the electric dispersion reactor (EDR), has been developed for the synthesis of ultrafine powders by emulsion techniques (7). This concept uses intense, pulsed, DC electric fields to cause jetting of a conducting liquid from a nozzle into a nonconducting organic liquid. As in the aforementioned case of electrostatic atomization in a gas, the jets decompose to form micron-sized droplets. An important difference is that reactants can be placed in the droplet phase and/or the continuous phase. Precipitation may be induced by the diffusion of re-

actant into the droplets or by varying the temperature. In addition to the increased flexibility of using various chemical methods, it appears that the EDR will display a decided advantage in both energy requirements and production rate because of the lower interfacial tension present in liquid-liquid systems versus liquid-gas systems.

The EDR has been used to synthesize micron size porous shells and dense spherical particles of single component and mixed hydrous oxide particles in organic solvents (7). This study will investigate the use of this device for producing ultrafine yttrium-barium-copper hydrous oxide precursor powders.

EXPERIMENTAL

Figure 1 shows a schematic of the proposed EDR continuous flow system. The basic concept of the EDR involves the creation of aqueous/organic emulsions. A drop of an aqueous solution is formed from a nozzle into the region between electrodes on which is imposed an intense, pulsing, dc electric field (6 to 10 kV/cm and pulsing frequencies between 1 to 3 kHz). Under properly selected conditions the drops are shattered, creating suspension consisting of micron-size droplets in a close size range. This is used as the basis for a reactor, with some of the re-

---

†Research sponsored by the Office of Basic Energy Sciences, U.S. Department of Energy under contract DE-AC05-84OR21400 with Martin Marietta Energy Systems, Inc.

††Managed by Martin Marietta Energy Systems, Inc

CONTINUOUS EDR PROCESS

ORNL DWG 88A-1256R2

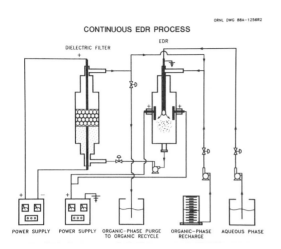

Figure 1. Configuration of the Electrical Dispersion Reactor

actants present in the continuous phase and some in the drops.

The operating mode involved placing the precipitation agent (ammonia) in the continuous organic phase (2-ethyl-1-hexanol or a mixture of 65 vol% 2-ethyl-1- hexanol, 35 vol% cyclohexane and 5 vol% ethanol) and the metal, in the form of metal salts (cupric chloride, yttrium nitrate, and barium nitrate) in the aqueous phase. For the purpose of this study, the system was operated in the batch mode without continuous-phase recycle.

After formation, the particles were removed by centrifugation and washed several times with ethanol. The particles were then dried for 24 hours. The particles were imaged before and after the washing and drying steps by scanning electron microscopy.

RESULTS AND DISCUSSION

Figures 2 and 3 are scanning electron micrographs of the particles. In Figure 2, the continuous phase was 2-ethyl-1-hexanol. Figure 3 shows particles produced when cyclohexane was added to decrease the level of $Ba(OH)_2$ solubility in the organic phase. In each case, the particles are spherical and the particle sizes range form 0.1- to 2-μm. Since all reactants are very soluble in the aqueous phase, diffusion is fast and precipitation occurs throughout the

Figure 2. Synthesis of Y-Ba-Cu (appox. 1:2:3) hydrous oxides by the dispersion of 0.975 M $Y(NO_3)_3$, 0.195 M $Ba(NO_3)_2$ and 0.2935 M $CuCl_2$ aqueous-phase into 0.17 M NH_3 in 2-ethyl 1-hexanol.

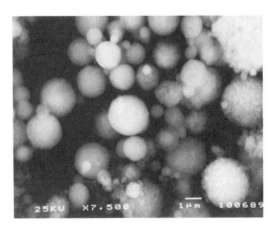

Figure 3. Synthesis of Y-Ba-Cu (approx. 1:2:3) hydrous oxides by the dispersion of the above aqueous solution (Fig. 2) into 0.16 M NH_3 in 65 vol% 2-ethyl 1-hexanol, 35 vol% cyclohexane and 5 vol% ethanol.

microdroplets. In most cases, the precipitated hydrous metal oxides remain in the aqueous phase since they are insoluble in the continuous organic phase. This is very attractive for the synthesis of mixed metal oxide particles because there is homogeneity on the submicron level.

The green phase is formed when the 1-2-3 powder is sintered. Subsequent chemical analysis shows that the powder is deficient in barium. The use of pure cyclohexane as the continuous phase (to eliminate barium in the continuous phase) results in a very low ammonia concentration in the continuous phase and poor reaction conditions. Subsequently, highly agglomerated and irregular shaped particles are formed. Other methods proposed (but not attempted in the present studies) include: (a) ammonia bubbled into the reactor to improve reaction conditions; and (b) all reactants are in the aqueous droplets and the ammonia source is hexamethylenetetramine (HMTA).

CONCLUSIONS

According to the afore-mentioned results, submicron- to micron-size hydrous metal oxide ceramic precursor particles can be formed by imposing intense, pulsing, dc electric fields on a conducting liquid drop suspended in a nonconducting liquid. Powders have been formed from the precipitation of aqueous metal salt solutions of copper, barium and yttrium. These powders were, however, deficient in barium; therefore, care must be taken in choosing a nonpolar solvent to eliminate the transport of barium hydroxide into the continuous phase. In a continuous processing scheme, this deficiency could be abated by allowing $Ba(OH)_2$ to build up to a steady level in the recycled organic phase; thereby, decreasing the driving force toward solubility in the organic phase. Future experimental work will involve: (a) developing methods to improve particle synthesis in nonpolar solvents; (b) larger-scale synthesis of powders in the EDR; and (c) comparing the sinterability, composition, and the ability of these powders to produce improved properties to precursors produced by other methods. Moreover, theoretical work is underway to shed light on the mechanisms of drop/particle formation from orifices in electric fields by finite/boundary element analysis (8).

ACKNOWLEDGMENTS

This research was sponsored by the Office of Basic Energy Sciences, U.S. Department of Energy. The authors would like to thank Ronald R. Brunson for his assistance.

LITERATURE CITED

1. Bowen, H., *Mat. Res. Soc. Symp. Proc.*, **24**, 1 (1984).

2. Martin, C.B., R.P. Kurosky, G.D. Maupin, C. Han, J. Javadpour and I.A. Askay, *Ceramic Trans.*, **12**, 99 (1990).

3. P. Odier, et al., *Ceramic Trans.*, **12**, 75 (1990).

4. Ingebrethsen, B.J., E. Matijevic, and R.E. Partch, *J. Colloid Interface Sci.*, **95**, 228 (1981).

5. Slamovich, E.B. and F.F. Lange, *Mat. Res. Soc. Proc.*, **121**, 257 (1988).

6. Osseo-Asare, K. and F.J. Aggiagada, *Ceramic Trans.*, **12**, 3 (1990).

7. Harris, M.T., T.C. Scott, O.A. Basaran, C.H. Byers, *Mat. Res. Soc. Symp. Proc.*, **180**, 853 (1990).

8. Harris, M.T. and O. A. Basaran, *Bull. Amer. Phys. Soc.*, **36**, 2960 (1991).

SYNTHESIS AND SCALEUP OF HIGH PURITY PRECURSORS TO SUPERCONDUCTING OXIDES

J. D. Connolly Jr., N. F. Levoy, and J. T. Schwartz ■ E. I. Du Pont de Nemours, Jackson Laboratory, Wilmington, DE

The development of a stable rare earth oxide, barium carbonate, and copper oxide powder precursor for use in fabricating powder and sintered shapes of $YBa_2Cu_3O_{(7-x)}$ superconducting oxide is described. The precursor is comprised of an intimate mixture of barium carbonate, copper oxide, and partially amorphous yttrium oxide. Two routes for the formation of this precursor powder are reported.

This precursor powder is advantageous in that it can be used in most processing environments without degradation of its properties. Once a shape has been fabricated, sintering the powder shape, with a proper oxygen anneal, will convert the powder to the superconducting oxide phase.

INTRODUCTION:

Since the discovery of the new class of metal oxide ceramics, which can conduct electricity with no resistance at temperatures above the boiling point of liquid nitrogen (77°K), a tremendous amount of research has been dedicated to adapting these materials for commercial applications. To accomplish this, these materials must be molded or formed into useful shapes, such as wires, films, or monolithic pieces.

A major disadvantage of superconducting oxides, such as $YBa_2Cu_3O_{(7-x)}$, is that they are unstable in water, acid and many other processing environments. Barium carbonate is a major impurity phase which forms when the so-called 1-2-3 powder is exposed to water and CO_2. This impurity hinders the sinterability of $YBa_2Cu_3O_{(7-x)}$ powder and degrades its properties [1]. In addition, the presence of barium carbonate in starting materials has been shown to inhibit rapid conversion to the desired superconducting phase [2,3]. For this reason aqueous processing of superconducting oxide powder is undesirable.

The objective of this program was to develop a precursor material which was stable in a variety of environments dictated by the forming techniques being considered. In this fashion a device would be formed using a stable powder, which would then convert readily to sintered superconducting oxide of the desired shape.

Two routes were developed for the formation of a stable rare earth, barium, and copper oxide precursor powder. Unlike the superconducting oxide, the precursor is stable in most processing environments. Upon heating to approximately 900°C, and with proper O_2 annealing, the precursor will readily convert to dense $YBa_2Cu_3O_{(7-x)}$ superconducting oxide. The homogeneity and crystal size are such that the precursor readily converts to the 1-2-3 oxide, despite the presence of barium carbonate. After formation, the precursor can be processed in a variety of ways including particle size reduction, formation of aqueous (basic or acidic) slurries, organic binder processes or dry powder forming techniques.

EXPERIMENTAL:

All metal salts (copper acetate monohydrate, yttrium acetate x-hydrate, and barium hydroxide octahydrate) were obtained

---

N. F. Levoy is now with Nuclear Metals, Inc., Concord, MA

commercially and were used without any further purification. Nanopure water refers to water which has been purified through a mixed bed ion exchange and organic/colloid removal column.

Route 1 Precursor:

As illustrated in Figure 1, copper acetate, yttrium acetate, and barium hydroxide were individually dissolved in Nanopure water. The solutions were then heated with agitation to a temperature of 75°C.

The yttrium solution was poured into the copper solution followed by the addition of the barium hydroxide solution over a period of 5 minutes while stirring. The mixture could be observed to turn from dark blue to cloudy green and then to brown during addition of the barium hydroxide, and a precipitate was observed to form. The mixture was then held at a temperature in the range of 70°C-80°C for 1 hour.

After 1 hour, the water was removed by spray drying to form a green-brown dry powder. This powder intermediate is deliquescent and was stored in a dry box until further processing into the precursor.

In order to form a stable precursor, the spray dried powder was calcined to at least 500°C to drive off excess water and to decompose the organics present in the powder. During calcination in air, barium carbonate, yttrium oxide, and copper oxide are formed.

Calcination of the intermediate powder was accomplished using a variety of furnace types (rotary or batch) capable of controlled temperature ramping to 500°C in an air atmosphere. The calcining material was contained in fused silica or quartz containers of size and shape appropriate for the furnace type and quantity of material to be calcined.

Decomposition of the spray dried intermediate resulted in the evolution of copious quantities of water and partial hydrocarbons. The sudden release of these materials caused the powder mass to undergo a transformation from a dry powder to a pliable body with a rubber-like consistency. This coupled with the evolution of large quantities of gas caused the reacting

mass to form large gas pockets, which in turn causes the bulk to swell to as much as ten times its original volume. The presence of the gas pockets are beneficial in the final product as they cause it to be friable and easily broken down. This was found to be beneficial for subsequent particle size reduction.

DTA traces of the initial spray dried powder showed sharp exotherms at 260°C, 310-325°C, and 365°C. Carefully controlled temperature ramps of from 1 - 4°C/min. were used to control the swelling. As the scale of superconductor precursor powder production increased, the capacity of the calcining furnaces was a consideration.

Route 2 Precursor:

As Route 1 was scaled several limitations became evident. The spray drying step accounted for as much as sixty-five percent of the processing time. The deliquescent nature of the spray dried intermediate made shelf life a problem. The solution route sometimes led to small batch to batch compositional differences . Additionally, the tremendous swelling of the calcining mass made it difficult to obtain the large furnace capacities required by larger scale operations.

In order to circumvent these problems, a second generation process was developed. In this route, as shown in Figure 2, the precursor powder was prepared by first dry blending the starting materials (copper acetate monohydrate, barium hydroxide octahydrate, and yttrium acetate x-hydrate) to form an intermediate blend. Next, the powder was loaded into a high shear blender. An equivalent amount of water by weight was added, and the blender was used to mix the ingredients, forming a slurry. The slurry was then subjected to additional high shear mixing for a period of 5-10 minutes, forming a blue foam. Care must be exercised at this point to prevent cavitation.

This blue foam can either be dried to form a glass-like solid, or heat treated prior to calcination. Heat treating, in contrast to drying, resulted in retention of water and of the foam consistency. Analysis of the precursor powder formed from either of these schemes indicated that heat treating is preferred.

Heat treating was accomplished by raising the temperature of the blue foam to 75°C for 40 minutes to an hour, or longer, with low speed agitation. This produced a dark green or khaki-brown colored foam, similar in color to the spray dried powder of route 1.

The foam was then calcined using the temperature ramps described above to produce the precursor powder. An advantage of using this route is that much of the expansion of the calcining precursor is eliminated in route 2, significantly increasing the furnace yield per batch. The homogeneity of the resulting precursor powder is also increased.

CHARACTERIZATION AND DISCUSSION:

Superconducting oxide (SCO) powders and their precursors, produced by the routes described above, were evaluated by several analytical techniques. X-ray fluorescence (XRF) and X-ray diffraction (XRD) were used to characterize bulk stoichiometry and crystal phase, electron microscopy techniques were used to characterize elemental homogeneity and microstructure.

XRF analysis of the precursor powder revealed a Ba/Y ratio of 1.99 and a Cu/Y ratio of 3.02. Following conversion of the precursor powder to the $YBa_2Cu_3O_{(7-x)}$, XRF analysis showed a Ba/Y ratio of 2.05 and a Cu/Y ratio of 3.06. These results indicate that, within experimental error, that each sample has the expected bulk stoichiometry for the $YBa_2Cu_3O_{(7-x)}$ compound. In addition, the bulk stoichiometry does not change significantly during conversion from the precursor powder to the SCO compound.

X-ray diffraction of the precursor powder indicates a mixture of crystalline $BaCO_3$, crystalline CuO and CuO_2, crystalline Y_2O_3, and an amorphous phase (Figure 3e). The amorphous phase has been shown by analytical electron microscopy to be Y rich. X-ray diffraction of the superconducting oxide powder produced by conversion of the precursor powder shows that the $YBa_2Cu_3O_{(7-x)}$ compound is the major phase, with varying degrees of phase purity depending on conversion conditions (Figure 4e).

As viewed in the scanning electron microscope (SEM), particles of the precursor powder are irregular in shape, with a size typically in the range of 1 to 30 microns (Figure 3a). Each particle is an aggregate of many finer particles. The nature of these aggregates is further revealed by higher resolution examination in the transmission electron microscope (TEM). This shows that the finer particles are a mixture of crystallites and amorphous components, typically in the size range of 40 to 150 nm. All crystallites observed are irregularly shaped, with rounded edges.

During conversion, the precursor powder sinters together to form larger crystallites. As viewed in the SEM these crystallites appear irregular in shape and have smooth surfaces (Figure 4a). Examination by TEM revealed that a majority of the particles observed were single crystals, in the range of 0.3 to 2.5 microns.

SEM with EDS (energy dispersive X-ray spectroscopy), was used to examine the distribution of elements in powder particles for both the precursor and SCO samples. Individual maps for Y, Ba, and Cu were generated for both precursor and SCO samples. As can be seen by comparing the elemental maps of Figure 3b,c,d with the corresponding SCO maps of Figure 4b,c,d, the chemical homogeneity of the particles increases with conversion to the SCO phase.

CONCLUSIONS:

A stable precursor for use in fabricating powder and sintered shapes of $YBa_2Cu_3O_{(7-x)}$ superconducting oxide has been developed. The precursor is comprised of an intimate mixture of barium carbonate, copper oxide, and partially amorphous yttrium oxide. The homogeneity and microcrystallinity are such that it readily converts to the $YBa_2Cu_3O_{(7-x)}$ superconducting oxide, despite the presence of barium carbonate.

LITERATURE CITED:

1. H.S. Horowitz, R.K. Bordia, R.B. Flippen, R.E. Johnson and U.Chowdhry, Material Research Bulletin **Vol 23**, p. 821 (1988).

2. B. Bender, L. Toth, J.R. Spann, et al, Advanced Ceramic Materials, **2, 3B**, 506 (1987).

3. K.G. Frase, D.R. Clarke, Advanced Ceramic Materials, **2, 3B**, 295 (1987).

4. R.S Roth, K.L. Davis, J.R. Dennis, Advanced Ceramic Materials, **2, 3B**, 303 (1987).

5. M.J. Cima, W. E. Rhine, "Powder Processing for Microstructural Control in Ceramic Superconductors", Ceramic Processing Research Report Laboratory Report #82, Mass. Inst. of Technol., Boston, (1987).

6. H.S. Horowitz, R.K. Bordia, C.C. Torardi, K.J. Morrissey, M.A. Subramanian, E.M. McCarron, J.B. Michel, T.R. Askew, R.B. Flippen, J.D. Bolt and U. Chowdhry, Solid State Ionics, **32/33**, 1087 (1989).

METAL ACETATE/HYDROXIDE ROUTE TO
PRECURSOR POWDER FOR 1-2-3 SUPERCONDUCTORS

METAL ACETATE/HYDROXIDE ROUTE TO
PRECURSOR POWDER FOR 1-2-3 SUPERCONDUCTORS

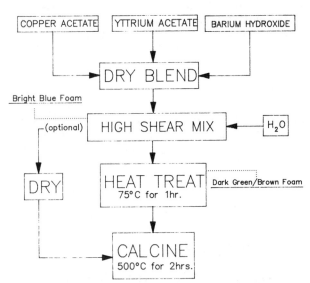

Figure 2.   Metal acetate/hydroxide route to precursor powder for 1-2-3 superconductors.

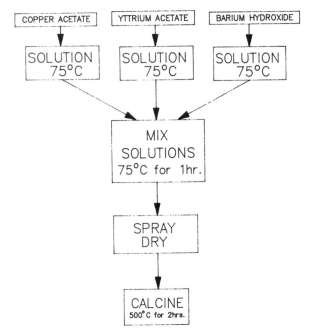

Figure 1.   Metal acetate/hydroxide route to precursor powder for 1-2-3 superconductors.

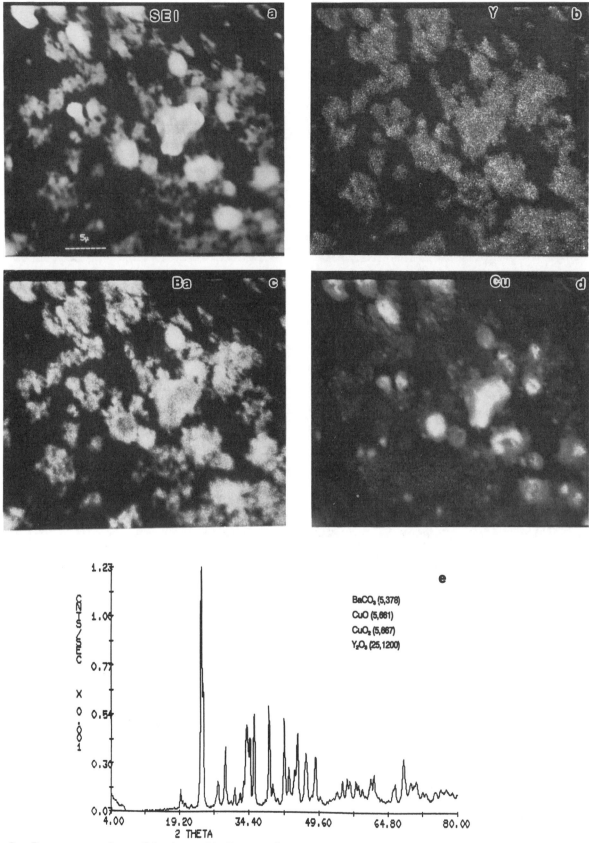

Figure 3.   Precursor powder particles imaged in the scanning electron microscope. (a) Secondary electron image, (b) X-ray elemental map for Y, (c) Ba and (d) Cu. (e) Corresponding X-ray diffraction pattern.

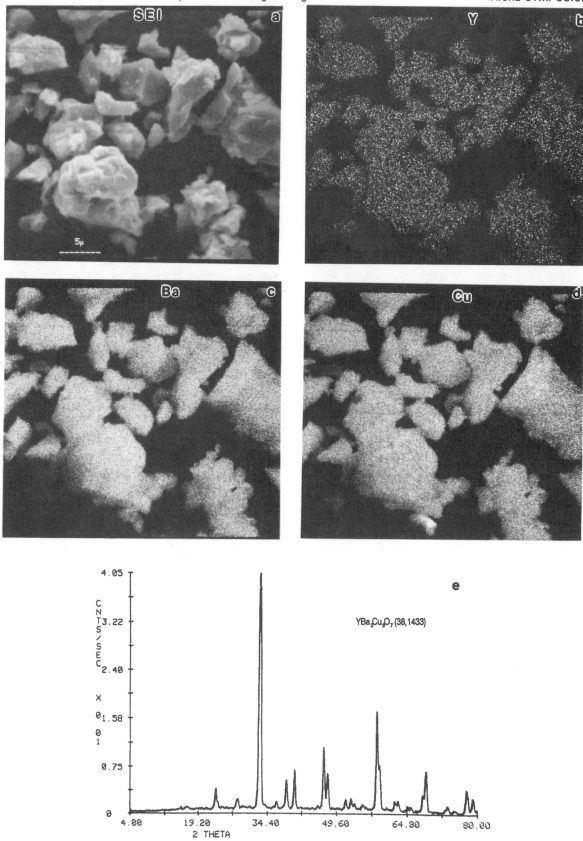

Figure 4.   Superconducting oxide powder particles prepared by conversion of precursor. (a) Secondary electron image, (b) X-ray elemental map for Y, (c) Ba, and (d) Cu. (e) Corresponding X-ray diffraction pattern.

FABRICATION AND WIRE EXTRUSION OF CERAMIC SUPERCONDUCTORS

R. B. Poeppel, U. Balachandran, J. P. Singh, J. T. Dusek, J. J. Picciolo, S. E. Dorris, M. T. Lanagan, K. C. Goretta, C. A. Youngdahl, and J. R. Hull ■ Materials and Components Technology Division, Argonne National Laboratory, Argonne, IL 60439

Many applications of high-temperature superconductors (HTSs) will depend on the ability to fabricate these materials into long lengths with suitable electrical and mechanical properties maintained over the entire length. The program described in this paper is focused on improvement of the relevant material properties of HTSs and on development of fabrication methods that can be transferred to industry for production of commercial conductors. Our research has resulted in advances in fabrication methods that improve the performance of long lengths of poly-crystalline HTS wires and tapes. We have examined the Y-Ba-Cu-O (YBCO), Bi-Sr-Ca-Cu-O (BSCCO), and Tl-Ba-Ca-Cu-O (TBCCO) classes of HTSs. Significant results from our research and work by contemporaries are reported in the various sections of the paper.

LOW-PRESSURE CALCINATION

A common method to synthesize HTS powder uses solid-state calcination and carbonates for one or more of the precursor powders. Our experience with this method is that calcination under low oxygen pressure improves the chemical and phase purity of YBCO and BSCCO

and greatly decreases the processing time required to achieve good HTS properties. Low-pressure calcination is beneficial in improving the kinetics of CO_2 removal and lowering the equilibrium temperatures in some cases, so that lower temperatures may be used; this results in a smaller particle size [1,2]. Reduced pressure has also proven beneficial in certain sintering procedures [3].

During calcination of the YBCO precursor, the desired perovskite phase is formed by the decomposition of $BaCO_3$ and the subsequent reaction among the three constituent oxides. CO_2 released by decomposition of $BaCO_3$ can, however, react with YBCO to form $BaCO_3$, Y_2O_3, CuO, and $Y_2Cu_2O_5$, depending on temperature. We have developed a synthesis route to obtain phase-pure orthorhombic YBCO powders at 800°C in flowing O_2 at reduced pressure [4,5]. We have optimized the heating rates and oxygen flow rates to synthesize large batches (1.2 kg) of powder. The powders produced by this reduced-pressure calcination technique were found to be pure by X-ray diffraction, differential thermal analysis, scanning electron microscopy (SEM), and transmission electron microscopy (TEM). In addition, magnetic susceptibility and T_c and J_c measurements indicated that superconducting qualities were good.

Large batches of phase-pure orthorhombic $YBa_2Cu_4O_y$ (124) have also been synthesized by the low-pressure calcination technique [1]. 124 has a more thermally stable oxygen content than 123 up to ≈820°C. Intragrain J_c (at 70 K) was ≈100 kA/cm^2 in zero field and 2 kA/cm^2 at 4 T. 124 decomposes into 123 and CuO; we have carefully controlled this decomposition and obtained uniformly distributed, fine precipitates (50-250 Å) of CuO in a 123 matrix. These fine precipitates will act as flux-pinning sites. The transition temperature is raised to 92 K from ≈70 K by the decomposition. Intragrain J_c was measured to be ≈1MA/cm^2 at 59 K in a magnetic field of 0.1 T.

Properties of YBCO superconductors depend strongly on sintering temperature and atmosphere. CO_2 is a common atmospheric contaminant. Also, CO_2 will evolve during the heating of YBCO that contains organics (binders, dispersants, solvents, etc., or residual carbonates left in the powder). Therefore, we have investigated the effect of CO_2-containing atmospheres during sintering of superconductors [6]. J_c decreased as the CO_2 content in the sintering atmosphere was increased. High-resolution electron microscopy showed two types of grain boundary phases: about 10% of the boundaries contained nonsuperconducting tetragonal phases. YBCO decomposed completely into several phases as the CO_2 content was increased to about 5% [7].

Reduced oxygen partial pressure has also been used in sintering YBCO samples for application experiments. Sintering pellets at ≈885°C in 1% O_2/balance N_2, followed by annealing in O_2 at ≈450°C gave the best results. The J_c in zero field at 77 K is ≈1000 A/cm^2 [3]. We have used this

technique to make current leads capable of carrying more than 2000 A [8]. Using the vacuum-calcined powder and sintering in reduced oxygen partial pressure (pO_2), we have fabricated a prototype coil by joining 15 rings. This coil, with an air core, had the capability of producing a 2.5 mT field.

$Bi_2Sr_2CaCu_2O_x$ (T_c ≈85 K) of very good phase purity has recently been synthesized. The actual composition of the superconducting phase is Sr-deficient and slightly Bi- and Ca-excess. Use of low pressure synthesis and precisely adjusted compositions has enabled powder of improved phase purity to be synthesized. Because of inter-growths characteristic of the BSCCO system, claims of phase-pure compounds are reserved until sufficient atomic-resolution TEM evidence can be obtained. Bulk specimens have been formed and processed by sintering in the solid state.

Several superconducting compounds exist in the TBCCO system. Deviations from ideal stoichiometry are extensive, however, and phase-pure materials have not yet been synthesized. Work by several research groups has shown that nearly phase-pure $Tl_2Ba_2Ca_2Cu_3O_x$ can be formed if excess Tl is used. Our work has shown that compounds deficient in Tl yield equivalent phase purities and T_c values. The advantage of the Tl-deficient compositions is that evolution of Tl-containing species is greatly reduced [9].

Melting of Tl_2O and evolution of Tl oxides are the most important considerations in processing TBCCO superconductors. Many researchers begin with Tl-excess mixtures to compensate for Tl evolution at high temperatures [9]. Our efforts have focused on encapsulation to avoid Tl volatilization. This approach is necessary for fabrication because of the impracticality of altering formulations as volatilization changes with every geometry used. The procedure begins by forming oxide precursors of the Ca, Ba, and Cu species:

Ca_2CuO_3, Ba_2CuO_3, and $Ba_2Cu_3O_5$. Pure precursors can be made by calcining in air from 900 to 950°C. Encapsulation can be successful because no CO_2 is evolved during heating.

POWDER-IN-TUBE FABRICATION

HTSs processed in Ag tubes have reportedly produced J_c values at 77 K in zero applied field of about 24 kA/cm^2 in BSCCO [10] and greater than 10 kA/cm^2 in TBCCO [11]. J_c values for YBCO of about 3000-4000 A/cm^2 appear to be reproducible; higher J_c values have not been adequately confirmed. Advantages of powder-in-tube processing include obtaining high green densities, which obviates the need for high sintering temperatures; protection of the superconductor from atmospheric exposure; and possible stabilization of the superconductor by the metallic sheath. Powder-in-tube processing has received increasing attention due to its ability to produce relatively high J_c in very high magnetic fields at 4 K [12].

Research reported here has included swaging [13] and rolling operations [2,14]. For both YBCO and BSCCO, the results to date can be readily summarized. Swaging was done with large areal reductions per pass. Rolling was done with deformation limited to 10% reduction per pass. When larger reductions were attempted, tensile stresses induced transverse cracks, and low J_c values are the resulted. Proper heat treatment yielded excellent, low-resistance YBCO/Ag interfaces. The BSCCO/Ag interfaces had higher resistance, and work is underway to modify heat-treatment times and atmospheres in order to improve the interfaces.

The microstructures developed to date have been only modestly textured. T_c and J_c values have been slightly higher than those of pressed pellets. The highest 77-K zero-field J_c for YBCO has reached 3500 A/cm^2 [15]. Advantages of powder-in-tube processing appear to be the ease with which long continuous lengths can be formed and the possibility of stabilization through the metallic sheath. Because this method of wire fabrication is highly directional, favorable texturing may be possible. Current work is focusing on obtaining greater extents of particle alignment. The two approaches are (1) use of powders with higher aspect ratios and (2) processing with an intermediate heat treatment to increase net plasticity and allow for favorable grain growth.

PLASTIC EXTRUSION

Plastic extrusion is a versatile process for forming long continuous lengths of HTS wire that are very flexible in the green state [16]. Because organic materials are used for plasticizers, binders, dispersants, and solvents, during extrusion, control of CO_2 evolution during the burnoff of the organics [4] and annealing [7] stages is as important as in the calcining stage. The presence of O_3 during burnoff of organics from extruded YBCO wires resulted in measurable increases in J_c; however, the presence of O_3 during the annealing stage had no noticeable effect [17].

Extrusions of $YBa_2Cu_3O_x$ have been coated with an insulator of Y_2BaCuO_5 and cofired to produce multilayer coils capable of several hundred ampere-turns of superconducting magnetomotive force [5,18] and used to power a small electric motor [19]. The best results were obtained when the coils were fired under low oxygen pressure. A 5-layer, 75-turn, 211-coated coil obtained 7.3 mT at 77 K with a J_c of 150 A/cm^2 over the entire coil. When an iron core was added, the maximum field obtained in the superconducting state was 33 mT at 77 K and 42 mT at 73 K.

A new technique is being tried for fabricating multilayer coils. With this technique, YBCO is extruded in the precursor form, and the 211 coating is applied as a precursor. That is, in the green state, the coil contains only uncalcined materials. The precursor wire is converted into YBCO, and the precursor coating into 211, when the coil is fired at reduced pressure. This technique combines the calcination and sintering steps of fabrication and could possibly avoid a problem with precalcined YBCO, i.e., its decomposition through contact with atmospheric contaminants such as CO_2 and H_2O. A single-layer coil has been made by firing a coil of YBCO precursor, and the results were promising: a J_c of ≈ 350 A/cm^2. It remains to be seen, however, if distinct layers of YBCO and 211 will form during calcination of the precursor coil.

MECHANICAL STRENGTH

The mechanical strength of wires and coils has been improved without degradation in critical current density by adding appropriate amounts of Ag powder to the HTS powder [20,21]. The Ag places the HTS crystals under compression and retards microcracking [22]. The combination of the above methods, plus several mechanical techniques to align crystallites (e.g., plastic extrusion, doctor-blade tape casting, and powder-in-tube processing), has resulted in continuous improvement of mechanical properties and critical current densities. Addition of Ag powder has also been used to improve the mechanical strength of YBCO coils [5,23].

As part of our effort to improve mechanical properties, extruded wires were heat-treated in a 100% flowing oxygen environment at temperatures of 925-950°C for different lengths of time (10-20 h) to obtain dense microstructures with small grains and improved mechanical and superconducting properties. As expected, density decreased with decreasing sintering temperature and time. Specifically, the density decreased from ≈98% to 91% of theoretical when the sintering temperature decreased from 950 to 925°C. Correspondingly, grain size decreased from ≈23 mm to 6 mm. The reduction in grain

size resulted in a corresponding increase in room-temperature strength from 83 to 141 MPa despite the small decrease in density. This observation is consistent with the well-known inverse dependence of strength on grain size. The increase in strength is due to a decrease in undesirable internal residual tensile stress with decreasing grain size. The decrease in residual stress helps reduce the failure-causing critical flaw size. Further reduction in sintering temperature to 910°C resulted in an additional decrease in grain size to ≈3 mm. But the decrease in grain size was accompanied by a much reduced density (≈79%).

To achieve a further decrease in grain size while maintaining high density, an effort has been initiated to sinter YBCO specimens at low pO_2 [3]. Initial results indicate that high densities can be achieved at low temperatures while maintaining fine-grained microstructures. Strengths as high as 200 MPa have been achieved with densities of 91% theoretical, an average grain size of 3-5 mm, and a J_c of ≈400 A/cm^2. The strength was also measured at liquid nitrogen temperature to confirm structural reliability at low temperatures; the strength was observed to be ≈10% higher than at room temperature. This is believed to be due to suppression of subcritical growth of the inherent failure-causing flaws, because crack growth is a thermally activated process.

A principal disadvantage of the perovskite superconductors for most envisioned applications is their brittleness. The fracture toughness of pure YBCO is about 1 MPa(m)$^{1/2}$. Partially stabilized ZrO_2 is often used to increase the fracture toughness of ceramic materials. Tetragonal ZrO_2 particles improve the fracture toughness of YBCO [24]. For some processing procedures, toughness was as high as 4.5 MPa(m)$^{1/2}$. As are most other compounds, ZrO_2 is chemically incompatible with YBCO, and ZrO_2 additions of several percent will destroy the superconducting properties, mainly through the formation of $BaZrO_3$. Recently, it was found that coating the ZrO_2 with Y_2BaCuO_5 minimizes the reaction between ZrO_2 and YBCO. A sol-gel coating process is presently the preferred method for coating ZrO_2 particles.

DIRECTIONAL SOLIDIFICATION

One of the greatest impediments for HTSs at present is the relatively low J_c compared with the more conventional superconductors such as NbTi and Nb_3Sn. A number of directional solidification techniques have been employed to obtain significant J_c values in magnetic fields of several teslas, and recently it has been demonstrated that at least some of these techniques are amenable to continuous processing [25]. So far, all directional solidification processes are too slow to have much commercial potential. However, the high J_c values that are obtainable make this process an active area of research.

Directional solidification work reported here was done in collaboration with the University of Notre Dame in what is called a zone-melt texturing process. In this process, wires are extruded and sintered as described previously. These wires are then zone-melted to develop highly textured and homogeneous microstructures [26]. Transport J_c values of more than 8 kA/cm^2 was measured in a field of 1 T at 77 K. Pulsed current measurement showed a transport J_c above 40 kA/cm^2 at 77 K in zero field. The magnetization J_c measured at 77 K reached 100 kA/cm^2 at 1 T [27]. The zero-field magnetization J_c at 77 K was greater than 2 MA/cm^2.

YBCO/Ag samples were also zone-melted; their microstructures contain large Ag and 211 precipitates within YBCO grains [28]. Transport J_c values significantly degraded when the Ag content exceeded 5 vol.%. The results are in contrast to sintered samples in which no degradation in J_c has been observed up to 15 vol.% Ag [22].

CONCLUSIONS

The creation of commercially usable HTS wire requires advances in a number of technologies related to mechanical, chemical, and electrical properties. Although considerable progress has been achieved since the discovery of HTSs, conductor properties are still suitable for only a limited number of applications. However, the rate of progress is encouraging, and if this rate continues, the development of HTS wires that operate at 77 K in commercial applications appears to be an obtainable goal in the near future.

ACKNOWLEDGMENTS

This work was supported by the U.S. Department of Energy, Office of Utility Technologies, Conservation and Renewable Energy, under Contract W-31-109-Eng-38. We are indebted to our many collaborators at ANL, other national laboratories, industry, and academia who have contributed greatly to the continuing success of our program.

LITERATURE CITED

1. Balachandran, U., M. E. Biznek, G. W. Tomlins, B. W. Veal and R. B. Poeppel, *Physica C*, **165**, 335 (1990).

2. Balachandran, U., M. J. McGuire, K. C. Goretta, C. A. Youngdahl, D. Shi, R. B. Poeppel and S. Danyluk, "Microstructure and Electrical Properties of Bulk High-T_c Superconductors," in *Superconductivity and Applications*, Kwok, H. S. (Ed.), Plenum, New York (1990).

3. Chen, N., D. Shi and K. C. Goretta, *J. Appl. Phys.*, **66**, 2485 (1989).

4. Balachandran, U., R. B. Poeppel, J. E. Emerson, S. A. Johnson,

M. T. Lanagan, C. A. Youngdahl, D. Shi, K. C. Goretta and N. G. Eror, *Mater. Lett.*, **8**, 454 (1989).

5. Dorris, S. E., J. T. Dusek, J. J. Picciolo, R. A. Russell, J. P. Singh and R. B. Poeppel, *Int. Conf. Elect. Mach.*, **1**, 7 (1990).

6. Gao, Y., Y. Li, K. L. Merkle, J. N. Mundy, C. Zhang, U. Balachandran and R. B. Poeppel, *Mater. Lett.*, **9**, 347 (1990).

7. Gao, Y., K. L. Merkle, C. Zhang, U. Balachandran and R. B. Poeppel, *J. Mater. Res.*, **5**, 1363 (1990).

8. Wu, J. L., J. T. Dederer, P. W. Eckels, S. K. Singh, J. R. Hull, R. B. Poeppel, C. A. Youngdahl, J. P. Singh, M. T. Lanagan and U. Balachandran, *IEEE Trans. Magn.*, **27**, 1861 (1991).

9. Goretta, K. C., D. Shi, B. Malecki, M. C. Hash and I. Bloom, *Supercond. Sci. Technol.*, **2**, 192 (1989).

10. Yamada, Y., K. Jikihara, T. Hasebe, T. Yanagiya, S. Yasuhara, M. Ishihara, T. Asano and Y. Tanaka, *Jpn. J. Appl. Phys.*, **29**, L456 (1990).

11. Torii, Y., H. Kugai, H. Takei and K. Tada, *Jpn. J. Appl. Phys.*, **29**, L952 (1990).

12. Heine, K., J. Tenbrink and M. Thöner, *Appl. Phys. Lett.*, **55**, 2441 (1989).

13. Shi, D. and K. C. Goretta, *Mater. Lett.*, **7**, 428 (1989).

14. Goretta, K. C., M. J. McGuire, A. Brandstädter, J. P. Singh, R. B. Poeppel, A. J. Schultz and J. L. Routbort, "Toughened $YBa_2Cu_3O_x/ZrO_2$ Composites," in *High Temperature Superconducting Compounds II*, Whang, S. H. (Ed.),

TMS Publications, Warrendale, PA (1990).

15. Shi, D., M. Xu, J. G. Chen, A. Umezawa, S. G. Lanan, D. Miller and K. C. Goretta, *Mater. Lett.*, **9**, 1 (1989).

16. Lanagan, M. T., R. B. Poeppel, J. P. Singh, D. I. dos Santos, J. K. Lumpp, U. Balachandran, J. T. Dusek and K. C. Goretta, *J.-Less-Common Met.*, **149**, 305 (1989).

17. Dorris, S. E., K. C. Goretta, B. Hajyousif, U. Balachandran, J. T. Dusek, D. Shi and R. B. Poeppel, *Ceram. Trans.*, **13**, 381 (1990).

18. Leu, H. J., J. P. Singh, S. E. Dorris and R. B. Poeppel, *Supercond. Sci. Technol.*, **2**, 311 (1989).

19. Edmonds, J. S., H. E. Jordan, J. D. Edick and R. F. Schiferl, *Int. Conf. Elect. Mach.*, **1**, 1 (1990).

20. Singh, J. P., H. J. Leu, R. B. Poeppel, E. Van Voorhess, G. T. Goudey, K. Winsley and D. Shi, *J. Appl. Phys.*, **66**, 3154 (1989).

21. Singh, J. P., D. Shi and D. W. Capone II, *Appl. Phys. Lett.*, **53**, 237 (1988).

22. Kupperman, D. S., J. P. Singh, J. Faber Jr. and R. L. Hitterman, *J. Appl. Phys.*, **66**, 3396 (1989).

23. Dorris, S. E., M. T. Lanagan, D. M. Moffatt, H. J. Leu, C. A. Youngdahl, U. Balachandran, A. Cazzato, D. E. Bloomberg and K. C. Goretta, *Jpn. J. Appl. Phys.*, **28**, L1415 (1989).

24. Goretta, K. C., O. D. Lacy, U. Balachandran, D. Shi and J. L. Routbort, *J. Mater. Sci. Lett.*, **9**, 380 (1990).

25. Meng, R. L., C. Kinalidis, Y. Y. Sun, L. Gao, Y. K. Tao, P. H. Hor and C. W. Chu, *Nature*, **345**, 326 (1990).

26. McGinn, P. J., W. Chen and M. A. Black, *Physica C*, **161**, 198 (1989).

27. Shi, D., H. Krishnan, J. M. Hong, D. Miller, M. Xu, P. J. McGinn, W. Chen, J. G. Chen, M. M. Fang, U. Welp, M. T. Lanagan, K. C. Goretta, J. T. Dusek, J. J. Picciolo and U. Balachandran, *J. Appl. Phys.*, **68**, 228 (1990).

28. McGinn, P., N. Zhu, W. Chen, M. Lanagan and U. Balachandran, *Physica C*, **167**, 343 (1990).

HIGH TEMPERATURE SUPERCONDUCTORS: CHALLENGES FOR A NEW TECHNOLOGY

T. E. Schlesinger ■ Department of Electrical and Computer Engineering, Carnegie Mellon University, Pittsburgh, PA 15213

The hurdles facing high temperature superconductors before they find general application in electronic devices and systems is discussed. It is argued that while many materials and processing problems are being addressed and overcome these may not ultimately determine whether or not these superconductors are employed widely in electronic systems.

There is no doubt that the discovery of high temperature superconductors [1] is a tremendously exciting and interesting development both in the fields of materials science and solid state physics. The unexpected observation of superconductivity above 77 K will, without doubt, lead to greater insights and understanding of fundamental interactions in the solid state and may even lead to the discovery of other interesting materials and physical effects. Along with these developments there has been a great deal of speculation as to the ultimate application of these materials in an engineering environment. While some of the requirements for the ultimate application of these materials are noted here, this paper is not meant as a review of the technical accomplishments that have been made towards the ultimate application of these materials. Rather, this paper points out the obstacles that high temperature superconductors must overcome before they might be incorporated in engineering systems. As will be discussed below, the obstacles to the use of these materials on a reasonably large scale may have nothing to do with their intrinsic material or physical limitations. The discussion presented here will be focussed on the use of high temperature superconductors for electronics though it would be expected that what is true of the electronics industry is true of other industries as well.

A review of the relative merits of GaAs versus silicon for electronic applications may serve as a basis for the discussion of high temperature superconductors in this area. In the electronics industry there is a well known saying that "GaAs is the material of the future and will always remain so". The intrinsically higher electron mobilities in GaAs have led many people to argue that at some stage GaAs digital integrated circuits would begin to displace those based on silicon technology. Indeed GaAs has been studied in great detail now for many years and the technology for producing fairly sophisticated systems and electronic devices based on this material (and associated III-V compounds) has been developed. Nonetheless, in most areas, GaAs has not displaced silicon technology to any great extent with some notable and important exceptions. As device dimensions have been reduced, as designs and packaging have been improved, the performance of silicon ICs have met the requirements of new generations of electronics. The advantage that GaAs might offer over silicon has been outweighed by the investment that would be required to switch to this technology. There are two important examples, however, where this last statement is not accurate and where GaAs (or more generally III-V materials) are definitely of great importance. These are in optical devices and electronic systems operating at frequencies above 2 GHz [2]. In the area of opto-electronics for optical communications, optical data storage systems, and in applications such as local area networks, microwave receivers, and so on that are moving into the frequency range above 2 GHz (where there is available bandwidth) III-V devices are predominant. The reason for this is simple. Silicon technology, as it stands now, cannot produce efficient light emitting devices nor can it easily produce circuits operating above 2 GHz. The latter point must be qualified with the observation that the development of germanium-silicon devices may allow this material system to extend the operating frequency of silicon based circuits beyond this limit [3]. Thus GaAs has succeeded in certain areas not because it has displaced silicon but because it did not have to compete with silicon in the first place. Improvements in silicon technology, in other words, which have not required radical changes in device and circuit fabrication technologies have generally succeeded over improvements that might be gained by employing GaAs but which require large scale investments and changes in technology.

We are thus led to ask, in light of these observations, what is the status of high temperature superconductors? It would appear that until an application area is identified where high T_c superconductors can provide devices that cannot be reproduced with existing technologies, even at somewhat lower levels of performance, these materials will have difficulty displacing existing technologies. It would seem therefore, that a more likely shorter term approach to the use of high temperature superconductors would require developing the technology necessary to make these materials compatible with existing electronic systems. Our own research efforts in this area have been pursued with this in mind.

To use high temperature superconductors in electronics, therefore, not only requires the ability to produce thin films with high T_c (the temperature where the material becomes superconducting) and high J_c (the maximum current density at any given temperature the superconductor can support before losing its superconducting properties) but to do so by means that are commonly employed in device fabrication facilities and on substrates that are commonly used for electronic applications. We can consider each of these points in turn.

The choice of substrate, being dictated by the applications area rather than the material requirements of the superconductors, appears limited to a small number of materials. For most electronic applications this means developing the ability to deposit these materials on silicon or insulators such as silicon dioxide. For microwave device applications alumina, sapphire, and a few other materials also appear to be candidate substrates. Our own work as well as the work of others has been aimed at depositing $YBa_2Cu_3O_{7-d}$ on silicon substrates. This is complicated by the diffusion of silicon into the superconductor as well as the diffusion of Ba into the silicon but can be overcome through the use of buffer layers such as yttria stabilized zirconium oxide and others [4]. The deposition technique to be employed must also be familiar to the processing engineer. This favors deposition methods such as single target on-axis sputtering and chemical vapor deposition. While other techniques such as off axis sputtering, laser ablation, multilayer evaporation, multi-target sputtering are being used to deposit these materials and in fact have probably produced the highest quality films to date [5] most of these techniques suffer from limitations in a production environment. For example, it is difficult to imagine laser ablation being used for very large area substrates though deposition rates can be very high [6], deposition rates for off-axis sputtering are low (a typical value may be on the order of 500 Å/hr [7]) and compositional uniformity over very large areas is difficult to achieve. By comparison on-axis sputtering is more easily scaled to larger areas and CVD techniques allow for deposition on multiple wafers with very good temperature uniformity and control. Indeed, our own work on single target on-axis sputtering has shown that even in a very small system it is possible to achieve compositionaly uniform depositions on silicon substrates, with deposition rates of about 1500 Å/hr, and superconducting transitions with in-situ processing [8] (heated substrates and oxygen anneals). Deposition rates are important to the process engineer for the following reasons. In a typical sputtering system used to deposit aluminum, say one micron thick, in integrated circuit technology the actual deposition time is small compared to the time to load wafers into the system, pump down, and so forth. If the deposition rate per wafer becomes comparable to the time necessary for preparation then a deposition process offering higher deposition rates (such as CVD or on-axis sputtering) begins to look more attractive in a fabrication facility than do other deposition methods.

The high temperature superconductors are layered materials in which high critical currents are achieved for current transport in directions normal to the crystallographic c-axis [9]. Thus it is necessary to deposit films with the c-axis oriented such that the high critical current directions are in the desired direction. This usually means having the c-axis normal to the substrate and this has been accomplished by various deposition techniques. It has also become clear that it is necessary to have low angle grain boundaries between a-b planes for high critical current densities and this too has been accomplished [10].

While there is much to be done in the way of optimizing these films and the associated deposition processes, work has also gone forward on developing processing technology to produce devices which incorporate these high temperature superconductors. There have been successful demonstrations of lithographic techniques to produce narrow lines and the deposition of multilayers of both insulators and metals and in general the technology necessary to produce devices. The superconducting layers have been shown to still maintain the desired properties even after the various processing procedures. Thus, in some sense, all the elements of a technology that can employ the high temperature superconductors has at least been demonstrated in the laboratory [11]. Indeed, in terms of discrete devices for certain applications of these materials one already sees commercial offerings.

The general use of these materials in electronics however faces a hurdle which is akin to the the ones discussed in the beginning of this paper and which has little to do with the superconductors per se. That is, even with T_c of 90 K or T_c of 120 K, in the Thallium based compounds, one can still only hope, at this point to operate devices at 77 K since superconducting devices generally must be operated at 50 - 75% of T_c. Thus for general applicability to electronics one is faced with the necessity of operating the entire electronic system at cryogenic temperatures. This is possible, indeed, work in low temperature electronics has preceded high temperature superconductors for the improvements in performance that semiconductor devices realize at low temperatures. The obstacles that prevent the general acceptance of low temperature electronics, however, stand in the way of the use of high temperature superconductors also [12]. Even in this application area one has to carefully weigh the relative advantages that might be gained by employing high temperature superconductors and introducing additional complexities to the fabrication processes versus employing more traditional materials. An example of this is the use of high temperature superconductors for interconnects. It might appear that with the zero resistance that these materials offer circuit speeds would be increased by the reduction of RC time constants. However, if one considers interconnects of aluminum versus those of high temperature superconductors at 77 K it is not at all clear that there is any advantage to be gained in terms of speed [13].

Thus, while ultimately one may expect high temperature superconductors to find applications in electronic systems and while various niche applications of high temperature superconductors may already exist the general use of these materials in electronics is not limited simply by the material or processing problems that must be overcome. Rather, the high temperature superconductors face the same obstacles that any new technology has in trying to displace an existing technology. The existing technology is mature and often can be improved to the point that the economics of moving to the new technology stands in the way of its acceptance. Ultimately it may be these issues more than any others that determine the applications of high temperature superconductors.

ACKNOWLEDGEMENT: The work on high temperature superconductors was carried out over many years with A.K. Stamper, D.W. Greve, M. Migliuolo, J.W. Lee, and D.D. Stancil. Work in this area has been supported by the National Science Foundation through a Presidential Young Investigator Award and by the Ben Franklin Partnership Program of the Commonwealth of Pennsylvania.

LITERATURE CITED:

1. Bednorz, J.G. and K.A. Muller, *Z. Phys. B* **64** 189(1986).

2. Watson, G.F. "Gigahertz nets spur gallium arsenide ICs", in *The Institute* Torrero, Edward (Ed.), IEEE Inc, New York (1991).

3. Patton, G.L., J.H. Comfort, B.S. Myerson, E.F. Crabbe, G.J. Scilla, E. De Fresart, J.M.C. Stork, J.Y.-C. Sun, D.L. Harame, and J.N. Burghartz, *IEEE Electron Dev. Lett.* 57, 1034(1990).

4. Mogro-Campero, A. Supercond. Sci. Technol. **3**, 155(1990).

5. Fork, D.K., D.B. Fenner, R.W. Barton, J.M. Phillips, G.A.N. Connell, J.B. Boyce, and T.H. Geballe, *Appl. Phys. Lett.* **57**, 1161(1990).

6. Wu, X.D. ,R.E. Muenchausen, S. Fottyn, R.C. Estler, R.C. Dye, A.R. Garcia, N.S. Nogar, P. England, R. Ramesh, D.M. Hwang, T.S. Ravi, C.C. Chang, T. Venkatesan, X.X. Xi, Q. Li, and A. Inam, *Appl. Phys. Lett.* **57**, 523(1990).

7. Newman, N., K. Char, S.M. Garrison, R.W. Barton, R.C. Taber, C.B. Eom, T.H. Geballe, and B. Wilkens, *Appl. Phys. Lett.* **57**, 520(1990).

8. Stamper, A.K., D.W. Greve, and T.E. Schlesinger, **J. Vac. Sci, Technol**. A 9, 2158(1991).

9. Tozer, S.W., A.W. Kleinsasser, T. Penney, D. Kaiser, and F. Holtzberg, *Phys. Rev. Lett.* **59**, 1768(1987).

10. Dimos, D., P. Chaudhari, J. Mannhart, and F.K. LeGoues, *Phys. Rev. Lett.* **61**, 219(1988).

11. Simon, R. *Physics Today* **44**, 64(1991).

12. Carlson, D.M., D.C. Sullivan, R.E. Bach, and D.R. Resnick, *IEEE Trans. Electron Devices* **36**, 1404(1989).

13. Watt, J.T., and J.D. Plummer "Effect of Interconnection Delay on Liquid Nitrogen Temperature CMOS Circuit Performance" presented at the 1987 International Electron Devices Meeting, IEEE, Washington D.C.

SYNTHESIS OF SUPERCONDUCTING COMPOSITE POWDERS VIA AEROSOL DECOMPOSITION

A.S. Gurav, T.L. Ward and T.T. Kodas ■ Department of Chemical and Nuclear Engineering, University of New Mexico, Albuquerque, NM 87131

and

J. Brynestad, D.M. Kroeger and H. Hsu ■ Oak Ridge National Laboratory, P.O. Box 2008, Oak Ridge, TN 37831

Superconducting composite powders of $YBa_2Cu_3O_{7-x}$/CuO (123/CuO), $YBa_2Cu_3O_{7-x}$/Y_2BaCuO_5 (123/211), $Yba_2Cu_3O_{7-x}$/Ag (123/Ag) and Bi-Pb-Sr-Ca-Cu-O (mixed phase) were synthesized by aerosol decomposition of nitrate solutions. Such composite powders are of interest for rapid conversion to another phase and for enhancing mechanical and transport properties (flux pinning) in sintered ceramics. The behavior of these powders during subsequent processing to form bulk superconducting ceramics is determined mainly by phase composition and grain size distribution. Experiments showed that the phase composition and grain size within the particles was determined primarily by the reactor temperature and residence time. For example, increasing the residence time and the reactor temperature resulted in elimination of impurities such as Ca_2PbO_4 and $(Bi,Pb)_2Sr_2CuO_x$ and increased the fraction of $(Bi,Pb)_2Sr_2Ca_2Cu_3O_x$ phase in the Bi-Pb-Sr-Ca-Cu-O system. The kinetics of phase formation for $YBa_2Cu_4O_8$ and $(Bi,Pb)_2Sr_2Ca_2Cu_3O_x$ (2223) were considerably enhanced by using aerosol decomposition. The 123/Ag powder contained silver in its elemental form and led to finer-grained sintered composites as compared to conventional solid-state reaction powders. Similarly, the 123/211 composite powders allowed formation of finer 211 inclusions (micron-sized) in the melt-textured 123 matrix than can be obtained by existing processes.

The discovery of high temperature superconductors such as Y-Ba-Cu-O [1], (Bi-Pb)-Sr-Ca-Cu-O [2,3] and Tl-Ba-Ca-Cu-O [4] has led to an explosion of research on these materials mainly because of their potential for applications ranging from microelectronic circuitry to large scale power generation and storage. However, several problems currently limit their applications; they have a relatively low capacity for carrying electrical current caused by the grain-boundary weak links and flux line movement, poor kinetics of formation for phases like $YBa_2Cu_4O_8$ (124) and $(Bi,Pb)_2Sr_2Ca_2Cu_3O_x$ (2223) and are inherently brittle. Some of these problems can be alleviated by utilizing powders which contain a homogeneous mixture of different components at a molecular level and consist of submicron-size particles with a narrow size distribution. Alternatively, composite powders of these superconductors with a fine scale dispersion of defects and secondary phases can enhance the kinetics of subsequent phase formation, mechanical properties and flux pinning characteristics.

Several approaches are available for preparation of powders used to form bulk superconducting ceramics. A highly promising approach is aerosol decomposition. The aerosol decomposition process involves generation of aerosol droplets from a solution containing dissolved metal salts such as nitrates, chlorides or acetates in the required stoichiometry. Co-decomposition of precursors within individual droplets takes place as the droplets are carried through a high-temperature reactor. This technique is capable of producing high-purity powders with either homogeneous chemical composition or controlled phase distributions (composite materials) and submicron particle size. Using this technique, it is easy to control stoichiometry and, in some cases, the powders can be processed further to obtain phase-pure materials in much shorter time compared to the conventional routes [5].

In this paper we describe synthesis of superconducting composite powders of $YBa_2Cu_3O_{7-x}$/CuO (123/CuO), $YBa_2Cu_3O_{7-x}$/ Y_2BaCuO_5 (123/211), $YBa_2Cu_3O_{7-x}$/Ag (123/Ag) and mixed Bi-Pb-Sr-Ca-Cu-O phases (BPSCCO). The dependence of the rate of formation and distribution of phases and the resulting superconducting properties on different processing parameters such as residence time and reactor temperature are reported. The results show that aerosol decomposition is a powerful route to formation of composite, high-purity submicron powders with controlled scales of phase segregation.

EXPERIMENTAL

A schematic of the powder generation apparatus is shown in Fig. 1. The carrier gas (humidified house air or commercial oxygen at 0.34 MPa) was passed through an aerosol generator and carried the aerosol droplets from the generator to the furnace. A TSI model 3076 atomizer was employed for aerosol generation from nitrate solutions. For some of the BPSCCO powders, a modified ultrasonic home humidifier (Pollenex SH55R) was used. The operation and characteristics of these atomizers have been described previously [5,6]. By varying the gas flow rate, the residence time could be varied between 3 to 50 s for reactor temperatures of 700 to 1000°C. A quartz or mullite reactor tube was used with an inner diameter of 8 cm and total length of 125 cm. The three zone furnace (Lindbergh) had independent PID controllers for the heated zones with a total heated length of 91 cm. The exit of the reactor tube led to a high efficiency nylon filter. The temperature of the filter holder was maintained at about 60°C using a heating tape to prevent water condensation, which can cause plugging and degradation of the powders.

An aqueous solution of metal nitrates with desired stoichiometry was prepared by reaction of Y_2O_3, CuO, $BaCO_3$ and Ag_2O (for Y-Ba-Cu-O) and Bi_2O_3, $SrCO_3$, CaO (or $CaCO_3$) and CuO (for BPSCCO) with HNO_3. Lead nitrate solution was prepared from $Pb(NO_3)_2$ powder. All chemicals (>99.99+%) were purchased from Aldrich. The low solubility of $Ba(NO_3)_2$ limited the solution concentration (0.025 to 0.035 M with respect to Y^{+3}) in the Y-Ba-Cu-O system. In the BPSCCO system, a pH of ~0.5 was necessary to prevent the precipitation of basic bismuth nitrate ($BiONO_3.H_2O$).

X-ray diffraction (Scintag USA, Cu Kα) was used to identify phases in the powders, providing information about phase purity and extent of reaction. Transmission Electron Microscopy (Philips CM 30 ST, operating at 200 kV) was used to examine particle and crystallite size. Superconducting transitions were measured by flux exclusion of an ac magnetic field (H$\approx$ 1 Gauss).

RESULTS AND DISCUSSION

Physico-chemical phenomena occurring during particle formation

The process of forming ceramic particles by aerosol decomposition can be viewed in the simplest case as the drying of droplets followed by solid and liquid phase chemical reactions within the particles as they flow through the reactor (Fig. 2) [7]. Evaporation of the solvent from the droplet is accompanied by nucleation of the metal salts in the particle to form one or more solid phases. This is followed by reaction of the precursors to form reaction intermediates which further diffuse and react to form the final product. Since all the species are initially mixed at the molecular level and the diffusion distances within the particles are short, the kinetics for the formation of multicomponent phases is considerably enhanced relative to solid state reaction routes. For example, orthorhombic $YBa_2Cu_3O_{7-x}$ could be produced within a residence time of 15 s at temperatures of 900-1000°C [5]. A problem encountered during aerosol decomposition of nitrate solutions is the formation of hollow or porous particles. However Kodas et al. [8] and Chadda et al. [5] have successfully demonstrated the production of solid particles by suitably increasing the reactor temperature and by reducing the solution concentration.

Phase Composition and Other Properties

Y-Ba-Cu-O System: Aerosol decomposition was used to produce $YBa_2Cu_3O_{7-x}/CuO$, $YBa_2Cu_3O_{7-x}/Y_2BaCuO_5$ and $YBa_2Cu_3O_{7-x}/Ag$ particles with controlled phase composition and microstructure.

Generation of 123/CuO powders was investigated to determine if the composite powders offered any advantages for conversion kinetics to the 124 phase [5]. The powders containing Y:Ba:Cu = 1:2:4 produced at 800-1000°C consisted primarily of 123, CuO and a small quantity of $Ba(NO_3)_2$ (Fig. 3). Crystallite size as observed by TEM was 25-50 nm and 100-250 nm, respectively, for powders synthesized at 800°C and 1000°C (residence time ~ 25 s) [7]. The fine-scale phase distribution in the powders synthesized at 800°C facilitated the conversion of this composite powder to $YBa_2Cu_4O_8$ (124) phase by heating at 750°C for 24 hr followed by 800°C for 24 hr at ambient pressure without using any catalysts. This rapid conversion of aerosol synthesized 123/CuO powders to 124 is encouraging when contrasted with the slow kinetics for formation of a high volume fraction of 124 phase that has been observed in previous studies. This slow transformation has in the past required the use of high oxygen pressures and catalysts like alkali nitrates or carbonates for 124 formation by conventional solid-state routes [9-12].

A motivation for generation of 124 phase material is for processing of 123 phase material with CuO inclusions. The 124 phase decomposes into 123 and CuO during heating to 920-950°C for a few minutes followed by rapid cooling to 700-750°C to retain the transformation-induced defects and to avoid reverse reaction. [13]. Such samples exhibit a superconducting transition at about 90 K and an intragrain value of J_c (Bean's model) increased by a factor of ten over the value for the typical 123 compound (from about 10^4 to about 10^5 A cm^{-2} at 77 K, H=0.9 T). The fine scale distribution of CuO in 123 matrix obtained in the as-synthesized aerosol powders may show similar advantages for increasing J_c.

The aerosol decomposition approach allowed formation of composite 123/211 particles which are useful as a precursor powder for melt-processing. The powders with Y:Ba:Cu = 1.6:2.3:3.3 produced at 900°C (residence time of 25 s) consisted of 123 phase, 211 phase and a small quantity of $Ba(NO_3)_2$ (Fig. 4). The trace of barium nitrate can be removed easily during subsequent processing [5]. A typical TEM revealed submicron solid particles of the 123/211 composite (Fig. 5). Thus, a suitable choice of stoichiometry in the precursor solution can lead to the formation of composite powders with the desired ratio of phases like 123 and 211. Studies on melt-texture processing of the 123/211 composites are reported elsewhere [14]. These powders led to finer 211 inclusions (micron-sized) as compared to the conventional solid state reacted composites showing 211 inclusions of 5 to 13 μm size. The dispersion of 211 particles reduced microcracks and segregation of impurities at the boundaries between 123 plates in the molten material.

Synthesis of 123/Ag powder allowed a new route to formation of ceramics with a fine-scale dispersion of Ag in 123 [18]. A dispersion of silver in a 123 matrix can improve the mechanical properties of 123 without chemically interacting with it [15-17]. The X-ray diffraction patterns of 123/Ag composite powder synthesized at temperatures between 900 and 950°C showed the presence of almost phase pure 123 and silver in its elemental form (Fig. 6). These powders exhibited a $T_c \approx$ 92 K in magnetic susceptibility measurements [18]. During sintering at 895°C, the silver grains remained fine and uniformly distributed and grew in size with increasing sintering time from ~1 μm after 2h to 3-7 μm after 60 h [18].

Bi-Pb-Sr-Ca-Cu-O System : A big advantage of aerosol decomposition for generation of BPSCCO powders is the ability to control the phase composition of the powders while providing a minimum amount of phase segregation. An understanding of what phases can be formed is vital for successful conversion of the precursor powders to the high-T_c 2223 phase. For this reason, the phase composition was studied as a function of process parameters. Examination of the phase compatibility for the Bi-Pb-Sr-Ca-Cu-O system showed that a large number of single or mixed metal oxides are stable at temperatures between 700°C and 1000°C [19-21]. Analysis of XRD data was carried out based on published literature [22-25]. Depending on the composition and process parameters like residence time and reactor temperature, the as-produced powders contained different proportions of phases such as $(Bi,Pb)_2Sr_2CuO_x$ (2201), $(Bi,Pb)_2Sr_2CaCu_2O_x$ (2212), $(Bi,Pb)_2Sr_2Ca_2Cu_3O_x$ (2223), Ca_2PbO_4, Sr_2PbO_4, Ca_2CuO_3 and CuO.

To study the effect of reactor temperature on phase formation, powders having the composition $Bi_{1.84}Pb_{0.37}Sr_{1.92}Ca_{2.0}Cu_{3.08}O_x$ were synthesized at 700°C and 800°C with a residence time of about 30 s. Fig.7(a,b) show the XRD patterns for these powders. Both powders had 2212 as the main phase. The powder synthesized at 700°C contained a large amount of 2201 phase and large quantities of Ca_2PbO_4 and Sr_2PbO_4. The powder synthesized at 800°C contained a small amount of 2201 and Sr_2PbO_4, while Ca_2PbO_4 was absent. The fraction of 2223 phase was higher than at 700°C. Since the as-synthesized powders consisted of a number of phases including some unidentifiable ones, quantitative estimation of the different phases could not be carried out. However, the above results clearly indicate that increasing the reactor temperature from 700°C to 800°C enhanced the reactions between 2201, 2212 and plumbate phases, leading to the formation of the 2223 phase. Reactor temperatures between 800°C and 850°C with sufficiently long residence time may lead to a considerable yield of 2223.

Fig. 8 shows another example of the effect of temperature on the phase content of powders produced in nitrogen with nominal composition $Bi_{1.4}Pb_{0.6}Sr_2Ca_2Cu_3O_x$ using a longer residence time of ~45 s. Again, the predominant phases are 2201 and 2212, and the higher reactor temperature led to essentially complete disappearance of Ca_2PbO_4. Though the residence time

was longer for these powders than for those in Fig. 7, no conclusions can be made about the residence time effect because of the different composition and gas atmosphere. Comparing Fig. 7 and Fig. 8 shows that the phase composition of as-produced powder in a complex mixed-oxide system such as this depends on temperature, residence time and the gas atmosphere. However, by fixing one or more of these variables, it appears that a consderable degree of control over the phase content may be achieved.

Unlike the compounds in the Y-Ba-Cu-O system which are point compounds existing at a particular ratio of cations, the phases in the Bi-Pb-Sr-Ca-Cu-O system can form over a range of compositions. Several groups have suggested regions in the ternary phase diagrams for optimum conversion to 2223 phase and better superconducting properties [26,27]. The phase composition plays an important role in the sequence of reactions and hence in the optimal conversion to the desired phases. Although the temperature and residence time used in this case are insufficient for complete reaction, the trend observed for the Bi/Pb ratio and the phases present agrees well with the results of Koyama et al. [26].

The aerosol-synthesized BPSCCO powders were heat treated to investigate the possibility of rapid conversion of the intermediate phases to the desired product (2223 phase), as observed during the conversion of 123/CuO to the 124 phase. The powders were sintered in air for 16 h at 850°C. The XRD pattern shown in Fig. 9 indicates a sizeable volume fraction (~79% as determined by relative intensities) of the 2223 phase in a sintered sample of $Bi_{1.8}Pb_{0.44}Sr_2Ca_{2.2}Cu_3O_x$. Susceptibility measurements on this sample showed a single transition with $T_c=110$ K (Fig. 10). The lack of a susceptibility transition around 80 K in spite of the presence of 2212 phase indicates that the 2223 phase development occurred such that the 2212 regions were shielded by the 2223 regions. Sintered samples of the nominal composition $Bi_{1.4}Pb_{0.6}Sr_2Ca_2Cu_3O_x$, showed a T_c of 78

K (Fig.10), indicative of the 2212 phase. The results on Pb-doped BSCCO material are encouraging considering the fact that the heating time used was short compared to times typically required for substantial 2223 formation. Silver sheathed wires made from the above aerosol synthesized BPSCCO powders have performed much better than those made from the solid state reacted powders and showed transport J_C values above 10^4 A cm^{-2} (at 77 K, zero field) [28].

SUMMARY

Superconducting composite powders of YBa$_2$Cu$_3$O$_{7-x}$/CuO (123/CuO), YBa$_2$Cu$_3$O$_{7-x}$/Y$_2$BaCuO$_5$ (123/211), YBa$_2$Cu$_3$O$_{7-x}$/Ag (123/Ag) and Bi-Pb-Sr-Ca-Cu-O (mixed phase) were synthesized by aerosol decomposition of nitrate solutions. The phase composition and grain size were primarily determined by the composition and processing parameters such as reactor temperature and residence time. The kinetics of phase formation for YBa$_2$Cu$_4$O$_8$ and (Bi,Pb)$_2$Sr$_2$Ca$_2$Cu$_3$O$_x$ were considerably enhanced by using powders synthesized via aerosol synthesis. The 123/211 composite powders allowed formation of micron-sized inclusions in the melt-textured 123 matrix. The aerosol decomposition route is highly promising for the synthesis of high-purity powders with either homogeneous chemical composition or controlled phase distributions (composite materials) and submicron size.

ACKNOWLEDGEMENT

The work was supported by NSF Grant # CTS 8908316 and by the U.S. Department of Energy, Assistant Secretary for Conservation and Renewable Energy, Office of Utility Technologies, Office of Energy Management/Advanced Utility Concepts - Superconductor Technology Program for Electric Energy Systems, under contract DE-AC05-84OR21400 with Martin Marietta Energy Systems, Inc.

LITERATURE CITED

1. Wu, M.K., Ashburn, J.R., Torng, C.J., Hor, P.H., Meng, R.L., Gao, L., Huang, Z.J., Wang, Y.Z. and Chu, C.W., Phys. Rev. Lett., 58, 908 (1987).

2. Maeda, H., Tanaka, Y., Fukutomi, M. and Asano, T., Jpn. J. Appl. Phys., 27, L209 (1988).

3. Sunshine, S.A., Siegriest, T., Schneemeyer, L.F., Murphy, D.W., Cava, R.J., Batlogg, B., van Dover, R.B., Fleming, R.M., Glarum, S.H., Nakahara, S., Farrow, R., Krajewski, J.J., Zahurak, S.M., Waszczak, J.V., Marshal, J.H., Marsh, P., Rupp, Jr, L.W. and Peck, W.F., Phys. Rev. B, 38, 893 (1988).

4. Sheng, Z.Z. and Herman, A.M., Nature, 332, 55 (1988).

5. Chadda, S., Ward, T.L., Carim, A., Kodas, T.T., Ott, K. and Kroeger, D., J. Aerosol Sci., 22, 5, 601 (1991).

6. Ward, T.L., Lyons, S.W., Kodas, T.T., Brynestad, J., Kroeger, M. and Hsu, H., Physica C (submitted).

7. Kodas, T.T., Adv. Mater., 1, 6, 180 (1989).

8. Kodas, T.T., Engler, E.M. and Lee, V.Y., J. Appl. Phys., 65, 2149 (1989).

9. Fischer, P., Karpinski, J., Kaldis, E., Jilek, E. and Rusiecki, S., Solid State Commun., 69, 531 (1989).

10. Morris, D.E., Asmer, N.G., Wei, T.Y.T., Sid, R.L., Nickel, J.H., Scott, J.S. and Post, J.E., Physica C, 162-164, 955 (1989).

11. Cava, R.J., Krajewski, J.J., Peck,Jr., W.F., Batlogg, B., Rupp, Jr., L.W., Fleming, R.M., James, A.C.W.P. and Marsh, P., Nature, 338, 328 (1989).

12. Buckley, R.G., Tallon, J.L., Pooke, D.M. and Presland, M.R., Physica C, 165, 391 (1990).

13. Jin, S. and Graebner, J.E., Mater. Sci. Engg., B7, 243 (1991).

14. Jin, S., Kammlott, G.W., Tiefel, T.H., Kodas, T.T., Ward, T.L. and Kroeger, D.M., Physica C, 181, 57 (1991).

15. Jin, S., Sherwood, R.C., Tiefel, T.M., van Dover, R.B. and Johnson, Jr., D.W., *Appl. Phys. Lett.*, **51**, 203 (1987).

16. Singh, J.P., Leu, H.J., Poeppel, R.B., van Voorhees, E., Goudey, G.T., Winshey, K. and Shi, D., *J. Appl. Phys.*, **66**, 3154 (1989).

17. Nishio, T., Itoh, Y., Ogasawara, F., Suganuma, M., Yamada, Y. and Mizutani, U., *J. Mater. Sci.*, **24**,3228 (1989).

18. Ward, T.L., Kodas, T.T., Carim, A.H., Kroeger, D.M. and Hsu, H., *J. Mater. Res.* (to be published, April 1992).

19. Lee, C.L., Chen, J.J., Wen, W.J., Perng, T.P., Wu, T.B., Chin, T.S., Liu, R.S. and Wu, P.T., *J. Mater. Res.*, **5**, 7, 1403 (1990).

20. Kitaguchi, H., Takada, J., Oda, K. and Miura, Y., *J. Mater. Res.*, **5**, 7, 1397 (1990).

21. Hettich, B., Bestgen, H. and Bock, J., *Adv. Mater.*, **3**, 6, 304 (1991).

22. Matheis, D.P. and Snyder, R.L., *Powder Diffraction*, **5**, 1, 8 (1990).

23. Shi., D., Tang, M., Vandervoort, K. and Claus, H., *Phys. Rev. B*, **39**, 13, 9091 (1989).

24. Tallon, J.L., Buckley, R.G., Gilberd, P.W. and Presland, M.R., *Physica C*, **158**, 247 (1989).

25. Onoda, M., Yamamoto, A., Muromachi, E.T. and Takekawa, S., *Jpn. J. Appl. Phys.*, **27**, 5, L833 (1988).

26. Koyama, S., Endo, U. and Kawai, T., *Jpn. J. Appl. Phys.*, **27**, 10, L1861 (1988).

27. Green, S.M., Mei, Y., Manzi, A.E., Luo, H.L., Ramesh, R. and Thomas, G., *J. Appl. Phys.*, **66**, 2, 728 (1989).

28. Kroeger, D.M., personal communication.

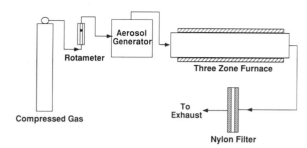

FIG. 1. Aerosol decomposition apparatus.

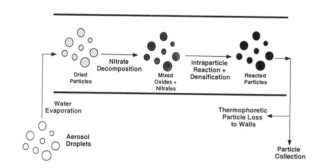

FIG. 2. Physico-chemical phenomena during particle formation.

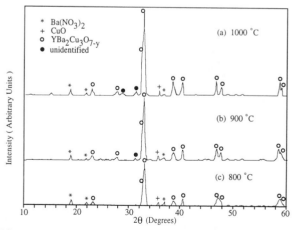

FIG. 3. XRD patterns (Cu, Kα) for powders with Y:Ba:Cu = 1:2:4, (a) 1000°C, (b) 900°C and (c) 800°C.

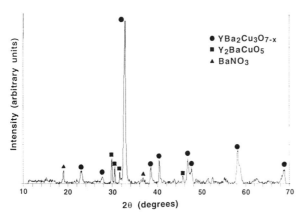

FIG. 4.   XRD pattern (Cu, Kα) for powders with Y:Ba:Cu = 1.6:2.3:3.3 produced at 900°C.

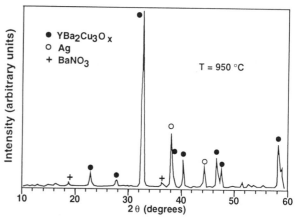

FIG. 6.   XRD pattern (Cu, Kα) for AgYBa$_2$Cu$_3$O$_{7-x}$ (123/Ag) powder produced at 950°C.

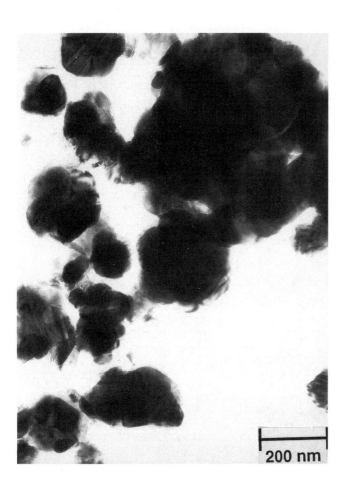

FIG. 5.   Transmission electron micrograph of as-produced 123/211 composite powder.

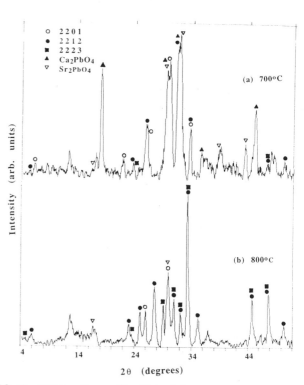

FIG. 7.   XRD patterns (Cu, Kα) of the as-produced powders having nominal composition Bi$_{1.84}$Pb$_{0.37}$Sr$_{1.92}$Ca$_{2.0}$Cu$_{3.08}$O$_x$ synthesized at (a) 700°C and (b) 800°C (residence time ~30 s).

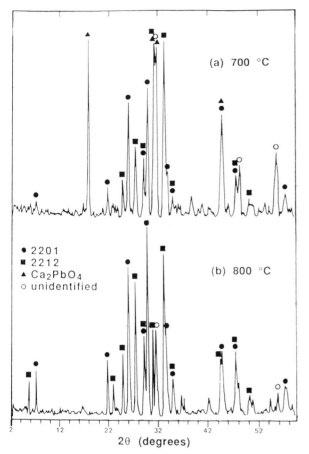

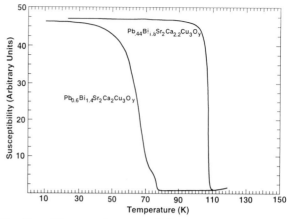

FIG. 10. AC magnetic susceptibility vs. temperature for (a) $Bi_{1.8}Pb_{0.44}Sr_2Ca_{2.2}Cu_3O_x$ and (b) $Bi_{1.4}Pb_{0.6}Sr_2Ca_2Cu_3O_x$ (sintered in air for 16 h at 850°C).

FIG. 8. XRD of as-produced powders having nominal composition $Bi_{1.4}Pb_{0.6}Sr_{2.0}Ca_{2.0}Cu_{3.0}O_x$ produced in nitrogen at (a) 700°C and (b) 800°C (residence time ~45 s).

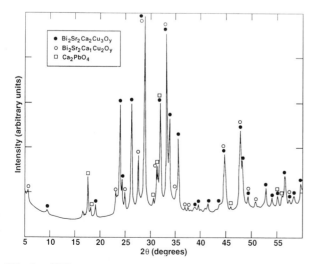

FIG. 9. XRD patterns of $Bi_{1.8}Pb_{0.44}Sr_2Ca_{2.2}Cu_3O_x$ powder sintered in air for 16 h at 850°C.

METAL-CERAMIC COMPOSITE SUPERCONDUCTING WIRES

A. Bhargava, M. A. Rodriguez and R. L. Snyder ■ Center for Advanced Ceramic Technology, Alfred University, Alfred, NY 14802

One of the more important applications of high temperature superconductors may involve the use of wires, fibers and rods. There have been numerous reports in literature about synthesis of wires and fibers, and include methods such as powder in tube [1]-[2], melt spinning [3]-[4], and melt growth [5]-[6].

The processing of superconductive forms by a glass to ceramic route has the potential of combining established continuous glass forming techniques to those of controlled heat treatment [7]. Thus, such a process has the potential for high strength and versatile shaped forms, including wires and rods. A variety of glass-ceramics have been synthesized that contain the $Y_1Ba_2Cu_3O_x$ and the $Bi_2Sr_2Ca_1Cu_2O_8$ (also known as the 2212) compound resulting in superconducting products [7]-[9].

The capability of bismuth to form glass in the presence of other modifiers may be exploited to synthesize glass-ceramics containing $Bi_2Sr_2Ca_1Cu_2O$ as the only crystalline phase [9]. This stoichiometry has now enabled the authors to form superconductive coatings on metallic wires without substantial loss in the flexibility of the parent wire. The process may be extended to other systems, such as those containing thallium, and to other forms such as tubes, rods and tapes.

EXPERIMENTAL PROCEDURES AND RESULTS

Reagent grade $SrCO_3$, CuO, Bi_2O_3, and $CaCO_3$ (Fisher Reagent Grade) were used in this study. A 100g batch was weighed out and intimately mixed by hand in a mortar and pestle. The equipment used to melt the batch consisted of a platinum crucible with a hole in it's bottom, suspended in a temperature controlled furnace. A thermocouple was placed very close to the surface of the crucible to monitor the temperature. The crucible was heated upto 1075°C and allowed to achieve thermal stability. About 10g of the batch was then transferred to the crucible. A melt was observed in approximately 5 minutes. The emphasis during melting was to melt at the lowest possible temperature for a minimum time period to avoid reaction of the batch with the crucible. This emphasis also yielded a higher viscosity thereby reducing the tendency of the undercooled liquid to crystallize. A platinum wire (approximately 100cm long) was pulled through the melt in a uniform motion allowing the melt to coat the wire. The drawing speed was approximately two feet per second.

A small piece of the coated wire was used for differential thermal analysis(DuPont 9900 Thermal Analyzer and DSC system) at a heating rate of 20°C/minute in an atmosphere of flowing nitrogen. Glass transition temperature, T_g was reported as the inflexion point in a step transition measurement as described by the Utility Program Manual published by E.I. DuPont de Nemour and Co. Inc.

Figure 1 shows a T_g at about 411oC and a T_x (where T_x is the crystallization peak temperature) at about 507°C, thereby showing presence of some amorphous phase. It is also observed that T_x-T_g, taken as a measure of glass stability [10] of the coating was about 96°C, which is significantly higher than that obtained (58°C) by a traditional melt-quench process [9]. The reason for this may be the higher cooling rate obtained during the coating process.

The coating on the wire was investigated by x-ray diffraction, XRD for possible crystalline phases by mounting the coated wire on a sample holder with an organic medium. A SIEMENS D500 diffractometer with Cu-Ka radiation and a diffracted beam monochromer was used. The surface of the coating was largely amorphous and showed presence of the crystalline phase $Bi_2Sr_2CuO_6$ (also known as 2201) oriented in the 001 direction. The coating was then removed from the wire, ground, and subjected to powder XRD analysis. This analysis confirmed the presence of an amorphous phase and the 2201 crystalline phase.

Samples of the coated wire, about 2cm long, were subjected to heat treatment at 850°C for a period of 32 hours. For heat treatment, the wire samples were suspended on platinum boats such that there was a free flow of sir. Heat treatment was conducted in an annealing furnace (Thermolyne Type 48000) at a heating rate of 10°C/minute. Surface and powder XRD patterns were obtained as described above and are shown in Figure 2(a)-(b). By comparing the calculated pattern of the 2212 compound with the observed, it may be concluded that the heat treated coatings contain the 2212 phase in high purity. The surface XRD analysis also indicate that the 2212 phase is strongly oriented in the 001 direction. The amorphous hump in the XRD pattern (a) arises from the mounting media and is therefore an instrumentation artifact. Though results from the 32 hour heat treated samples are quoted in this study, similar results are obtained on samples heat treated for lesser time periods, such as 2 hours.

Scanning electron microscopy, SEM (ETEC) was performed on the surface of the wire samples. The as-obtained sample showed a microstructure very typical of glass. The surface of the heat treated wires show crystallites of 2212 in the 1-10 micron range with significant orientation. Samples of the heat treated wire were also mounted in cross-section, polished to 1 micron and lightly etched with nitric acid. It was observed that the microstructure was dense with unifrom particle size in the 1 micron range.

Heat treated wires were also tested for resistivity using a four-point method [7] with an APD Cryogenics liquid helium cryostat and a Keithley 228 current source. Voltage drop was measured by a Keithley 196 model DMM voltmeter. The heat treated samples showed a superconducting onset temperature of about 90K and a zero resistance at about 77K (see figure 3). As an extension of this measurement, the current was increased till breakdown. The current which the sample was able to tolerate before breakdown was noted as the critical current. Critical current densities thus calculated were 10^2-10^3 amperes/cm^2 at 40K. Similar success was obtained with nichrome and silver wires with thicknesses ranging from 0.1mm to 1.0mm. diameter.

Using the same process, we also coated alumina and mullite rods and tubes of 1-2mm diameter. As before, the initially amorphous coating was devitrified by heat treatment as described earlier for the platinum composite. Though XRD results showed the crystalline phases to be 2201 and 2212, the composite did not display a superconducting transition in the liquid nitrogen range. Upon examining the samples under SEM, it was seen that there were large crystallites of 2201 randomly distributed, perhaps obliterating all current paths. It is also possible that there might be some aluminum incorporated in the structure of the 2212 that might be deleterious to the superconducting behavior.

Although we could observe no stress cracking in any of the samples, a thermal expansion mis-match could possible result in mechanical problems during use. Thermal expansion measurements were done on a vitreous and devitrified pieces of the 2212 composition. To obtain samples, part of the batch obtained earlier was melted at 1075°C for 10 minutes and quenched between aluminum plates. The glass thus formed was annealed at 400°C for 4 hours and allowed to cool to room temperature in about 6 hours. Samples were gently polished to obtain parallel sides and lengths of nearly 18mm. A sample of the parent glass was heat treated at 850°C for 32 hours to crystallize the 2212 phase. It was confirmed by XRD measurements that this sample was indeed the 2212 phase.

Four-point resistivity measurement indicated a Tc onset of 90K and a Jc of about 10^3 amperes/cm^2. The crystallized sample was also polished to have parallel edges and a length of about 18mm. Both the glass and the heat treated samples were measured for coefficient of linear thermal expansion, a, in the 200-400°C range, using an Orton horizontal vitreous silica dilatometer. For the amorphous precursor a is observed to be 12.2 x 10^{-6}/K while for the crystallized sample it is observed to be 12.0 x 10^{-6}/K. Therefore, we see good binding with platinum as a core metal, which has a of about 9.1 x 10^{-6}/K.

CONCLUSIONS

We have successfully made superconducting wires by a process of melt coating on a metal core and subsequent heat treatment. The superconducting onset temperature is about 90K and critical current density at 40K is in the 10^2-10^3 amperes/cm^2 range.

LITERATURE CITED

1. Heine, K., J. Tenbrink and M. Thoner, *Appl. Phys. Lett.*, **55**, 23, 2441 (1989).

2. Millington, J.S. and R.C. Sherwood, US Patent 4,952,554 (1990).

3. Ginley, D.S., E.L. Venturini, J.F. Kwak, M.A. Mitchell, and B. Morosin, *J. Appl. Phys.*, **67**, 6388 (1990).

4. Catania, P., N. Hovnanian, and L. Cot, *Mat. Res. Bull.*, **25**, 12, 1477 (1990).

5. He, Y., J. Zhang, A. He, J. Wang, and Y. Huo, *Superconductor Sc. and Tech.*, **4**, S1, S154 (1991).

6. Millington, J.S., and R.C. Sherwood, US Patent 5,011,823 (1991).

7. Nishio, M., K. Hayashi, Y. Nakai, K. Okhura   and K. Sawada, US Patent 4,973,574 (1990).

8. Bhargava, A., A.K. Varshneya and R.L. Snyder, *Mat. Lett.*, **8**, 1-2, 41 (1989).

9. Bhargava, A., A.K. Varshneya and R.L. Snyder, *Mat. Lett.*, **8**, 10, 425 (1989).

10. Zhao X., and S. Sakka, *J. Non-Cryst. Solids*, **95-96**, 487 (1987).

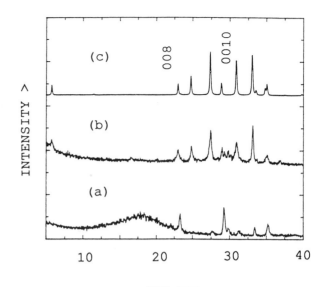

Figure 2.   XRD on (a) surface and (b) powder of coated samples compared to calculated pattern of 2212 (c).

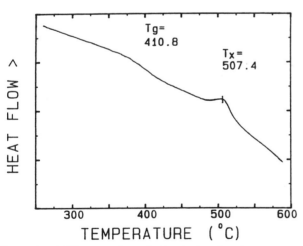

Figure 1.   DSC trace of as-obtained wire showing presence of amorphous phase in coating.

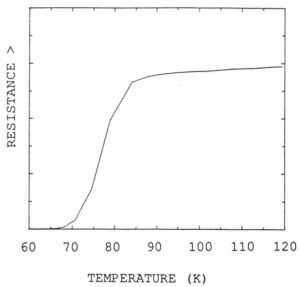

Figure 3.   Resistance versus temperature behavior of devitrified wire showing a T_c.

A CONTINUOUS COPRECIPITATION PROCESS FOR THE PRODUCTION OF 1-2-3 PRECURSORS

B. F. Allen, N. M. Faulk*, S.-C Lin, R. Semiat**, D. Luss, and J. T. Richardson ■ Department of Chemical Engineering, University of Houston, Houston, TX 77204-4792

A continuous process for the production of 1-2-3 superconductors by coprecipitation has been developed, based on a previous batch procedure. Inexpensive alkali reagents are used together with readily available equipment, and the process is easily scaled to higher production rates. Correct 1-2-3 stoichiometry is achieved with a low amount of alkali that is easily removed during calcination. The finished powder particle size is about one micrometer, and the final superconductor displays excellent properties.

Oxide superconductors were first made by mixing fine-grained solid powders and reacting them at high temperatures for periods of hours to days.[1,2] This method is effective on a small scale but is limited. First, $BaCO_3$, used together with Y_2O_3 and CuO, inhibits formation of the 1-2-3 phase, since decomposition does not occur below 1123K. Second, $BaCO_3$-CuO mixtures form eutectics that are difficult to reverse. Phase separation then restricts formation of the 1-2-3 phase. Third, uncalcined powders are larger than 25 micrometers, producing materials with poor microstructure and mechanical strength. Repeated grinding and high temperature calcination steps over extensive periods are necessary to overcome these problems. This poor control of particle size, morphology, homogeneity, and purity makes scale-up very difficult.

Improvement in product quality is possible with routes using solutions of soluble salts of 1-2-3 components. Coprecipitation, used extensively in large-scale production of powders, is an effective approach where good stoichiometry together with particle size control is necessary.[3,4,5,6] For commercial applications, the method should use inexpensive and common precipitating reagents and be adaptable to scaled-up, continuous unit operations. These requirements are also found in manufacturing of multicomponent catalysts,[7] in which preferred reagents are hydroxides and carbonates of sodium and potassium.

Sodium and potassium ions readily adsorb on the precipitates and are potentially harmful to superconductor materials. They must be removed by extensive washing during which there is a high risk of cation loss with departure from desired stoichiometry. Hydroxide precipitation has some advantages since carbon contamination is avoided. Hydroxides of copper and yttrium hydroxides form at pHs below 7 and barium hydroxide above 12, where other cations are soluble. Kini et al. [8] used a procedure in which 1-2-3 nitrate solution was neutralized to a pH of 7-8 with KOH or NaOH. Potassium carbonate solution was then added and the precipitate centrifuged and washed with water adjusted to a pH of 9.7 to avoid loss of barium. This technique introduces inhomogeneities into the precipitate. Addition of KOH to the

---

* Now at the Department of Chemical Engineering, Tennessee Technological University, Cookeville, TN 38505.
**Technion Israel Institute of Technology, Haifa 3200, Israel

nitrate solution may also result in formation of basic nitrates of copper, which later dissolve to give poor stoichiometry, and yttrium hydroxide precipitates early. Unless care is taken to stir the solutions well, addition of the carbonate to the nitrate produces uneven precipitation in the high pH regions of the mixture.

Bunker et al. achieved complete and simultaneous precipitation in a flow system, using both CO_2 and tetra-methylammonium hydroxide.[9] Extensive washing was avoided and they achieved good stoichiometry. This reagent is too expensive for an economical process. In previous reports from this laboratory, we have described a batch process for the coprecipitation of high purity 1-2-3 precursors with mixtures of common reagents such as sodium and potassium hydroxides or carbonates.[10,11]

Complete coprecipitation of 1-2-3 precursors was achieved by adding an aqueous solution of mixed nitrates rapidly to a mixture of sodium and potassium hydroxide and carbonate producing blue hydrogel. This intimate mixture of $Y(OH)_3$, $BaCO_3$ and $Cu(OH)_2$, was filtered, washed and reslurried. After a second wash by decantation, using sparged CO_2 to control the pH, the gel was dried and calcinated. The product was a homogeneous powder, with an average particle size of about one micron. Alkali impurity was minimized during the controlled washing step and residual amounts vaporized during calcining. The superconductor was free of alkali and exhibited excellent properties. The process scales easily to higher production rates and is inexpensive. Important parameters are: precipitation time; alkali ratios; volume, temperature and pH of the washing; and time of calcination.

Clearfield et al., [12] also using a batch approach, later found similar results, but these authors' did not address the question of alkali contamination or product homogeneity. Neither did they use the CO_2 sparge-washing, which we believe achieves best performance.

In this paper, we report an adaptation of our method to continuous operations and describe a unit for the production of two kilograms per day of precursor powder that yields one kilogram per day of superconductor. Emphasis is placed on the scale-up of critical parameters.

PROCESS DESCRIPTION

Adaption of batch processes to large-scale, continuous operations requires careful matching of batch conditions within the restrictions of continuous units. The optimized procedure for the batch process is as follows:

(1) A 1-2-3 mixture of nitrates is prepared between 0.1 and 0.6M at a pH of 3.3. The alkali solution is a 2M mixture of Na or K hydroxide and carbonate with the $OH^-/CO_3^=$ ratio between 4:1 and 1:4. The amount of carbonate is one to three times that required to precipitate the barium ions.

(2) The mixed nitrate solution is added rapidly with vigorous stirring to the alkali solution, both at room temperature. The pH decreases from 14 to 12.5 in three minutes, forming a blue precipitate which consists of an intimate mixture of yttrium hydroxide, barium carbonate and copper hydroxide. Longer times or higher temperatures cause a color change from light blue to grayish-brown.

(3) The suspension is filtered immediately and washed with distilled water, using 6 to 10 x 10^3 cm^3 of deionized water for each mole of cation. The filter cake is mixed with 15 to 30 x 10^3 cm^3 of water for each cation mole and washed by decantation for ten to thirty minutes while maintaining the pH from 7 to 9 by sparging CO_2, preventing cation loss and improving the washing by peptizing the precipitate. Following a second filtering, the mixture is washed and dried in air at 380K, during which

the color changes from light blue to gray.

Critical features of the process are (1) $OH^-/CO_3^=$ ratio range, (2) final pH between 12 and 13, (3) washing time, and (4) CO_2 sparging during the final wash. When these conditions are satisfied, satisfactory results occur. For example, cation stoichiometry remains 1-2-3 throughout and alkali compositions are 6 to 7 wt % after the first wash, 0.5 to 0.7 wt % after the second wash, and non-detectable after final calcination and annealing.

For continuous operations, a stirred tank reactor or precipitator is an effective substitution for the batch coprecipitator. Adjusted nitrate and alkali streams are pumped into the stirred vessel at rates that maintain a constant pH and provide the desired production rate. The degree of mixing and rate of precipitation depends on the size of the vessel, the type of impellor and the location of the inlet tubes.[13] In addition, the residence time affects extent of precipitation, stoichiometry and particle size.

We combined washing and filtering into a single operation using a continuous filter in which the initial wash and the CO_2-treated wash could be duplicated to achieve the desired alkali removal without sacrificing stoichiometry. These features were investigated in exploratory studies in which the effect of process conditions were examined. This led to an optimized process that successfully duplicated the results of the batch recipe while providing for scale-up and improved product quality control.

Compositions were measured with a Perkin Elmer ICP 5500. Precision of stoichiometry was determined within 1.5% and the limit of alkali detection was 2 ppm. Phase structure was determined with a Seimens 5000 Diffraction System, equipped with a position-sensitive detector. Particle size distributions (PSD) were measured

with an Horiba Particle Size Distribution Analyzer CAPA 700, and densities with a Quantachrome Mult Pycnometer.

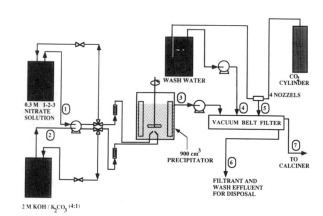

Figure 1.   The continuous process for 1-2-3 precursor pro-duction.

Figure 1 shows the process unit configuration and flow sheet. There are three parts to the process: the solution supply, stirred precipitator and the belt filter. The cation solution consists of a 0.3M 1-2-3 mixture of yttrium, barium and copper nitrate solutions and the alkali solution is a 2M mixture of KOH and K_2CO_3 with a ratio of 4:1 and a carbonate to cation ratio of 0.4. These stock solutions are stored in Nalgene tanks, from which a stream of each is pumped to the precipitator using a Cole-Palmer Master Flex Pump No. 7533-20 equipped with an Easy Load Pump Head No. 7518-10 to accommodate Streams 1 and 2. Flows are measured with two Gilmont Flow Meters, Size 4. The flow rate of each stream is adjusted by the pump to give a pH of 12.5 in the precipitator. For most operations, these rates are 167 and 68 $cm^3 \ min^{-1}$ for the cation and alkali solutions respectively. These flows are designed to yield approximately 300 grams of dry 1-2-3 precursor per hour. For a seven-hour run and a 50 weight percent loss on calcining, this corresponds to a production rate of about one kilogram per day of 1-2-3 superconductor. By operating for three shifts, this unit could produce over three kilograms per day.

Constructed of Plexiglass, the stirred tank precipitator is 10 cm in diameter and 15 cm in height and uses a LabMaster Model TS1515 mixer with a 6.4-cm turbine blade impellor and four baffles. The maximum tank volume is 900 cm^3. The impellor is positioned 3.5 cm above the bottom of the tank, and the two feed streams enter below the blade through 0.6 cm i.d. tubes with both ends located 2.5 cm above the bottom and bent at 45° with the tips 0.5 cm apart. This design ensures maximum turbulence to achieve rapid and simultaneous precipitation of the 1-2-3 components. A Cole-Palmer pH Meter Model No. 05986-25 monitors the pH continuously.

The outlet tube for Stream 3 is positioned to give whatever solution volume is desired, which, was usually 500 cm^3, and the product suspension is pumped with a second Master Flex pump at rates that maintain steady state conditions and a residence time of 2.1 minutes.

The effluent of Stream 3 emerges at the beginning of a vacuum belt filter, constructed in-house. This unit comprises a belt of Tetko 5-1550SK30 Hybrid Polypropylene, with an effective width of 6.5 cm, mounted on motor driven rollers to provide a filtration length of 83 cm and moving at 4 cm min^{-1}. A sealed space below the belt provides is evacuated to a vacuum of better than 40 kPa.

There are three sections to the filtering area. In Section No. 1 the slurry is distributed evenly on the belt surface and the excess liquid removed for a belt length of 10 cm. The filter cake thickness is about 0.25 cm. Distilled water, pumped from a container with a Master Flex Pump at a rate of 155 cm^3 min^{-1}(Stream 4), is forced over the filter cake for another 5 cm of belt movement. This corresponds to the initial wash in the batch procedure and uses 5 x 10^3 cm^3 of water per mole of cation, compared with 6-10 x 10^3 cm^3 per mole for the

batch process. Less water is necessary for continuous operation, which helps to prevent cation loss.

In Section No. 2, water saturated with CO_2 to a pH of 4 is sprayed over the cake for a distance of 23 cm at a rate of 250 cm^3 min^{-1}. The nozzles are Air Mist Gravity/Siphon Spray Nozzles Model No.156.521.16.03. This simulates the CO_2 sparging and washing of the batch recipe. Water is used at 18 x 10^3 cm^3 per mole which is about half that of the batch process. In the final section, the cake is dehydrated, removed from the belt with a scraper, and collected. Loss of 1-2-3 component is typically less than one percent.

The filter cake, the precursor for the 1-2-3 superconductor, must be dried, calcined and annealed for final production. We are currently exploring the use of a continuous rotary kiln for this stage of the process. Details and operating conditions of this unit will be given in a later report.[14]

Figure 2 shows the effect of time on the stoichiometry after the precipitator. The residence time was four minutes. Stoichiometry remains constant over a long period of time, which is an improvement over the batch process which showed cation loss after three minutes.

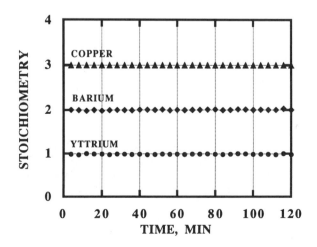

Figure 2. The effect of process time on 1-2-3 stoichiometry.

Figure 3 gives the potassium impurity level for the same samples. Although some variation is observed,

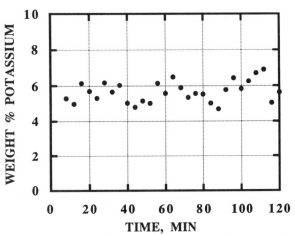

Figure 3.   The effect of process time on potassium concentration.

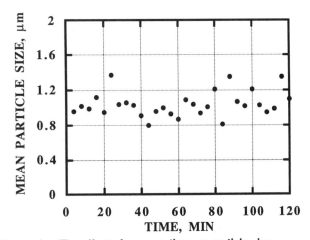

Figure 4.   The effect of process time on particle size.

the values fluctuate about a mean value of about 5.5 wt %, which is less than that obtained in the batch procedure, which suggests that the continuous mixing is giving better results. Figure 4 shows the effect on the mean particle size which fluctuates about a mean value of 1.1 microns. These results indicate that time has very little effect on critical product qualities.

Table 1 shows typical results from a production run.

TABLE 1
Typical Production Run Results

Stoichiometry:

Y	1.006
Ba	2.003
Cu	3.000

Density (wet), g cm^{-3}:	1.02
Density (dry), g cm^{-3}:	3.46
Potassium, wt %:	0.60
Particle Size, micrometers:	1.2

A sample of the filtration product was treated under laboratory conditions that were known to produce satisfactory products. The filtrate was first dried overnight at 383K in an oven and then calcined in flowing air with 20% excess oxygen. The temperature was raised at 20 K min^{-1} to 1193K, held for six hours and then cooled to room temperature. The final composition was $Y_{1.0104}Ba_{2.002}Cu_3O_{6.5}$ and structure analysis by X-ray diffraction showed that the sample consisted of 97-98% 1-2-3 phase with just a trace of 2-1-1. The crystallites were large and oriented in the c direction.

CONCLUSIONS

A continuous coprecipitation process has been developed duplicating the performance of a previous batch procedure while giving better results. Inexpensive reagents are used, with only commonly available equipment. Production rates of up to three kilograms per day are possible with the existing configuration, and scale-up to higher rates could be easily achieved.

ACKNOWLEDGMENTS

This work was supported by the Texas Center for Superconductivity at the University of Houston, under grants from the Defence Advanced Research Projects Agency and the State of Texas, and by the Advanced

Technology Program of Texas Higher Education Coordinating Board.

LITERATURE CITED

1. Wu, M. K., J. R. Ashburn, C. J.Torng, P. H.Hor, R. L.Meng, L. Gao, Z. J. Huang, Y. Q. Wang, and C. W. Chu, Phys. Rev. Lett. **58** 908 (1987).

2. Kawai,T.,M.Kanie,Jap.J.Appl. Phys., **26** L736 (1987).

3. Bender, B., L. Toth, J. R. Spann, S. Lawrence, J. Wallace, D. Lewis, M. Osofsky, W. Fuller, E. Skelton, S. Wolk, S. Qadri, and D. Gubser, Adv. Ceram. Mater., **2** (3B) 506 (1987).

4. Makoto, K., Jap. J. Appl.Phys., **26** L1821 (1987).

5. Dunn, B., C. T. Chu, L-W Zhou, J. R. Cooper, and G. Gruner, Adv.Ceram. Mat., **2** 343-352 (1987).

6. Wang, C. T., L. S. Lin, J. H. Lin, J. Y. Su, S. J. Yang, and S. E. Hsu, Mat. Res. Soc. Symp. Proc., **99** 257 (1988).

7. Richardson, J. T., _Principles of Catalyst Design_, Plenum Pub.Corp.,New York (1989).

8.Kini, A. M., U. Geiser, H. C. I. Kao, K. D. Carlson, H. H. Wang, M. R. Monaghan, and J. M. Williams, Inorg. Chem., **26** 1834 (1987).

9. Bunker,B. C.,J. A. Voight, and H.D. Doughty, _Superconducting Materials: Preparation,Properties and Processing_, (W. E. Hatfield and J. H. Miller,Ed.), M. Dekker, Inc., New York, 1988.

10. Morgan, D., M. Maric, D. Luss, and J. T. Richardson, "Preparation of 1-2-3 Superconductors from Hydroxide-Carbonate Coprecipitation", National AIChE Meeting, San Francisco CA, November 1989.

11. Morgan, D., M. Maric, D. Luss, and J. T. Richardson, J. Am. Ceram. Soc., **73**, 3557 (1990).

12. Clearfield, A., R. A. Mohan Ram, R-C Wang, and D. C. Dufner, Mat. Res. Bull., **25**, 923 (1990).

13.Tosun, G., "An Experimental Study of the Effect of Mixing on Particle Size Distribution in $BaSO_4$ Precipitation Reaction", Proceedings from the Sixth European Conference on Mixing, Pavia, Italy, May 1988, 161.

14. Luss, D., J. T. Richardson, H. G. K. Sundar, and S. Shelukar, to be published.

EFFECT OF AMBIENT ENVIRONMENT AND PRESSURE ON THE FORMATION OF $YBa_2Cu_3O_{6+x}$ FROM Y_2O_3, $BaCO_3$ and CuO

S. Sundaresan ■ Department of Chemical Engineering, Princeton University, Princeton, NJ 08544

The reaction pathway through which $YBa_2CU_3O_{6+x}$ (123) is formed from a precursor material containing finely mixed Y_2O_3, $BaCO_3$ and CuO prepared by a spray pyrolysis process has been studied. The precursor powder was loosely packed in cylindrical crucibles and heated for different lengths of time in N_2 and in O_2 at different temperatures and pressures. The samples were then layered and analyzed by XRD to assess the spatial variations in the composition. The tetragonal 123 phase was the primary reaction product that could by identified by XRD. At 700-750°C and atmospheric pressure, the formation of the 123 phase was found to occur at first in the outer layer of the packing and subsequently this front moved into the specimen, in both N_2 and O_2 environments. Experiments carried out at lower pressures revealed that a decrease in pressure resulted in an enhancement in the conversion. Similar trends were obtained with dry-pressed compacts of the precursor as well. This shows clearly that the rate of removal of CO_2 from the interior of the sample affects the progress of these reactions in a profound way.

Two binary mixtures $2BaCO_3 + 3CuO$ and $4BaCO_3 + Y_2O_3$ corresponding to the stoichimetry as in the 123 compound were prepared by a spray pyrolysis technique. Thermogravimetric analysis of these binary mixtures and the precursor for 123 revealed that the rate-limiting step in the formation of 123 is the reaction between $BaCO_3$ and CuO.

INTRODUCTION

The effect of processing environment on the formation of 123 superconductor from precursor compound has been a subject of several studies. Rha et al. [1] showed that the time required for the formation of the tetragonal phase of the 123 superconductor and the subsequent sintering process could be considerably reduced if processing were conduced in an oxygen-free environment. Horowitz et al. [2] reported that, when an inert environment was employed, a variety of sol-gel precursors can lead to formation of the 123 phase below 1000 K. Thomson et al. [3] have shown that although the fully developed 123 tetragonal phase is formed at temperature as low as 880 K in helium, this phase decomposes to Y_2BaCuO_5, $BaCu_2O_2$, and $BaCuO_2$, with the extent of decomposition depending on temperature. The presence of oxygen in the gaseous environment delayed this decomposition to much higher temperatures. In a detailed and systematic reaction sequencing study, Ruckenstein et al. [4] found that the rate-limiting step in the formation of 123 from a mixture of $BaCO_3$, CuO and Y_2O_3 was the decomposition of $BaCO_3$. Gadalla and Hegg [5] studied the kinetics of 123 formation using TGA and DTA analyses and concluded that the formation and decomposition of the 123 compound followed six overlapping steps that are all diffusion controlled. Lay [6] and Grader et al. [7] found that lowering the total pressure, i.e. vacuum processing, was highly beneficial to the formation of 123. Basu and Searcy [8] have shown that the decomposition temperature of $BaCO_3$ can be reduced by ~100°C at a low pressure of oxygen and CO_2.

It is known that the 123 superconductor can be formed as thin films under processing temperatures below 900 K [9,10], while much higher temperatures (~1200 K) are usually needed for synthesizing thick objects. The issue of scale up has been addressed by Ruckenstein et al. [4] who demonstrated that the conversion of precursors to 123 decreased appreciably when the sample size was increased.

The present study is also concerned with the effect of temperature, pressure and gas phase composition on the formation of 123 from precursors. The novelty of the present study lies in the type of results presented. We report extents of reaction measured at different distances away from the sample surface to bring forth the highly nonuniform pattern of chemical reaction.

EXPERIMENTAL PROCEDURE

Precursor to the 123 superconductor prepared from "flashed nitrates" was supplied by SSC Corporation (Seattle, WA). This preparation procedure involves a "flashed" carbothermal reduction of the nitrate salts to produce intimately mixed Y_2O_3, CuO and $BaCO_3$. A solution of yttrium, barium and copper nitrates at appropriate stoichiometric ratio and sucrose was spray dried and then rapidly pyrolyzed. The resulting powder was further heat-treated in air for one hour at 500°C. Precursor powder containing only yttrium and barium ($Y_2O_3 + 4BaCO_3$), and barium and copper ($2BaCO_3 + 3CuO$) were also prepared in the same manner. In what follows, we will refer to the ($Y_2O_3 + 4BaCO_3 + 6$ CuO), ($Y_2O_3 + 4BaCO_3$), ($2BaCO_3 + 3Cuo$) powders as precursor A, B and C respectively. Controlled heat treatments were carried out in the temperature range of 700-800°C, at different pressures in the range of 1-760 torr, with different bulk gas phase compositions (oxygen, nitrogen and a mixture containing oxygen and carbon dioxide at a molar ration of 40:1). The powder specimen to be heat treated was loaded into a cylindrical crucible (10mm diameter x 9mm height), while constantly tapping the crucible to encourage good packing. In this manner 1gm of precursor powder could be loaded into the crucible. This configuration was preferred for our studies for the following reasons: (a) there is a well-defined one-dimensional diffusional path for gas phase molecules inside the sample, which renders interpretation easy; (b) after heat treatment, the sample could be easily partitioned in six fractions consisting of material withdrawn from different depths from the free surface; (c) the initial packing density was low enough to avoid agglomeration of the powder upon heat treatment. The heat-treated specimen were analyzed by XRD.

The weight loss characteristics of the precursor samples were followed by TGA.

RESULTS AND DISCUSSION

The XRD patterns obtained with the precursor A powder and a sample obtained after heat treatment at 700°C for 6 hrs. in 760 torr N_2 are shown in Figure 1. The dominant peaks at 2θ values of about 24° and 33° for $BaCO_3$ and 123 respectively can be clearly seen. The intensities of these peaks (I_{BaCO3} and I_{123}) were used as quantitative measures of the amounts of these two species in our analysis. Then the ratio $I_{123}/(I_{123} + I_{BaCO3})$ is a measure of the extent of conversion of the precursor (A) to 123 during heat treatment.

Figures 2 and 3 present the variation of conversion with sample depth after a heat treatment at 700°C for 1 hr. and 6 hr. respectively. Results for several different N_2 pressures are shown in these figures. One can readily notice that, at a total pressure of 1 atm, there is a pronounced spatial variation in conversion, with conversion primarily restricted to a region near the external surface. Decreasing the pressure has a dramatic effect on the conversion.

Figures 4 and 5 show the effect of heat treatment temperature on the conversion at two different N_2 pressures. Increasing the temperature results in an increase in conversion. But at 760 torr N_2 pressure, the large spatial nonuniformity persists even at 750°C.

Figure 6 shows the results obtained with heat treatment at 700°C using different gaseous environment. The conversions obtained in an inert environment (N_2) are slightly higher than those obtained with oxygen. Note than when a small amount of CO_2 was added to the gas phase, the conversion was almost completely inhibited.

Figure 7 shows the TGA results obtained with precursor A powder at different gaseous environments. Note that the presence of CO_2 in the gas phase has dramatically delayed the onset of weight loss. The reaction proceeds slightly faster in an N_2 atmosphere than in oxygen, and this has been observed in earlier studies.

These results clearly demonstrate that the critical factor in the solid state reaction process is CO_2 in the gas phase. The presence of CO_2 in the gas phase exerts an inhibitory effect on the reaction. The variation of conversion with depth is then readily understood. The buildup of CO_2 in the interior of the sample hinders further reaction. Lowering the gas pressure decreases the partial pressure of CO_2, thereby facilitating the reaction. Increasing the temperature increases the equilibrium partial pressure of CO_2 above which little progress will come about in the solid state reaction. This then permits a higher conversion at a higher temperature.

A number of experiments were carried out using pressed pellets of precursor instead of tapped piles. The removal of CO_2 from the interior of specimen is more difficult in the case of pressed pellets than in tapped piles. So one would expect much lower conversions in the interior of pressed pellets than tapped piles, if CO_2 removal is a critical factor. This was indeed found to be the case. With pressed pellets (5 mm thick), conversion of the precursor was restricted to a very thin crust near the surface, when heat treatments of the sort described in figures 2 to 6 were carried out.

From the effect of CO_2 observed in our studies, one can readily conclude that the rate-limiting reaction in the formation of 123 must include $BaCO_3$. It can be easily demonstrated that the mechanism leading to the formation of 123 from precursor A does not involve the reaction

$$BaCO_3(s) = BaO(s) + CO_2(g).$$

From straightforward thermodynamic calculations, it can be shown that this reaction will proceed only at much higher temperatures than those used in the present study. Therefore, either Y_2O_3, CuO or $CU_2Y_2O_5$ must be facilitating the decomposition of the barium carbonate. Ruchkenstein it al.[4] have shown that the reaction between $BaCO_3$ and CuO leading to $BaCuO_2$ is an important step in the mechanism. To confirm this, we carried out TGA studies with precursors B and C. See figures 8 and 9. By comparing these figures

with figure 7, one can readily see that the key reaction is indeed the one between $BaCO_3$ and CuO:

$$BaCO_3(s) + CuO(s) = BaCUO_2(s) + CO_2(g)$$

Additional verification of this is provided in figure 10, where the weight-loss behaviors at a fixed temperature (670°C) obtained with two different $2BaCO_3 + 3CuO$ specimen are presented. The filled squares correspond to spray pyrolized precursor (C). The open squares correspond to a specimen prepared by mechanically mixing $BaCO_3$ and CuO powders. The intimacy of mixing is clearly better in the case of spray-pyrolyzed specimen, which indeed reacted faster.

SUMMARY

It is clearly demonstrated that the key step in the formation of 123 from a mixture of $BaCO_3$, CuO and Y_2O_3 is the reaction between $BaCO_3$ and CuO, confirming the finding of Ruckenstein et al.[4]. These authors reported a dramatic scale-up problem. We have established that an alteration of thermodynamic force caused by the buildup of CO_2 in the interior of the sample is the root cause for the scale-up problem. Evacuation facilitate CO_2 removal, thereby increasing the thermodynamic driving force.

ACKNOWLEDGEMENT

This work was performed at the University of Washington, Department of Materials Science and Engineering, where the author spent a sabbatical leave in 1988-89. The assistance of Dr. Ilhan A. Aksay, and Dr. C. Han of University of Washington are greatly appreciated.

LITERATURE CITED

1. Rha, J.J., K.J. Yoon, S.L. Kang, and D.N. Yoon, "Rapid Calcination and Sintering of $YBa_2Cu_3O_x$ Superconductor Powder Mixture in Inert Atmosphere," *J. Am. Ceram.* soc., 71, [7] C-328 - C-329 (1988).

2. Horowitz, H.S., S.J. McLain, A. W. Sleight, J.D. Druliner, P.L. Gai, M.J. VanKavelaar, J.L. Wagner, B.D. Biggs, and S. J. Poon, "Submicrometer Superconducting $YBa_2Cu_3O_{6+x}$ Particles

Route," *Science (Washington)*, 243, 66–69 (1989).

3. Thomson, W.J., H. Wang, D.B. Parkman, D.X. Li, M. Strasik, T.S. Luhman, C. Han, and I.A. Aksay, "Reaction Sequencing During Processing of the 123 Superconductor," *J. Am. Ceram. soc.*, 72 [10] 1977–79 (1989).

4. Ruckenstein, E., S. Narain, and N.L. Wu, "Reaction Pathways for the formation of the $YBa_2Cu_3O_{7-x}$ Compound," *J. Mater. Res.*, 4 [2] 267–72 (1989).

5. Gadalla, A., and T. Hegg, "Kinetcis and Reaction Mechanisms for Formation and Decomposition of $Ba_2YCu_3O_x$, "*Thermoschim. Acta*, 145, 149–63 (1989).

6. Lay, K.W., "Formation of Yttrium Barium Cuprate Powder at Low Temperatures," *J. Am. Ceram. soc.*, 72 [4] 696–8 (1989).

7. Grader, G.S., P.K. Gallagher, and D.A. Fleming, "Effect of Starting Particle Size and Vacuum Processing on the $YBa_2Cu_3O_x$ Phase Formation," *Chem. Mater.*, 1 [6] 665–8 (1989).

8. Basu, T.K., and A.W. Searcy, "Kinetics and Thermodynamics of Decompostion of Barium Carbonate," *J. Chem. soc., Faraday Trans.1*, 72 [9] 1889–95 (1976).

9. Wu, X.D., A. Inam, T. Venkatesan, C.C. Chang, E.W. Chase, P. Barboux, J.M. Tarascon, and B. Wilkins, "Low–Temperature Preparation of High–T_c Superconducting Thin Films," *Appl. Phys. Lelt*, 52 [9] 754–56 (1988).

10. Miura, T., Y. Terashima, M. Sagoi, and K. Kubo, "Low–Temperature Preparation of As–Sputtered Superconducting YBaCuO Thin Films by Magnetron sputtering, "*Jpn. J. Appl. Phys.*, 27 [7] L1260–L1261 (1988).

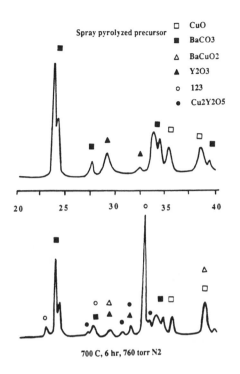

Figure 1: XRD patterns of precursor A before and after heat treatment at 700°C.

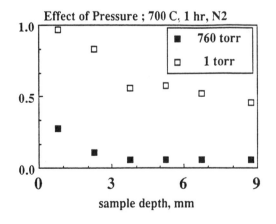

Figure 2: $\dfrac{I_{123}}{I_{123} + I_{BaCO_3}}$ vs. sample depth.

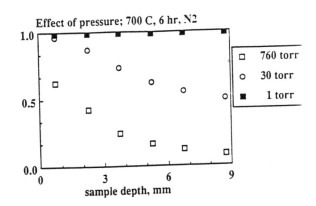

Figure 3: $\dfrac{I_{123}}{I_{123} + I_{BaCO_3}}$ vs. sample depth.

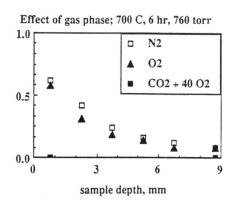

Figure 6: $\dfrac{I_{123}}{I_{123} + I_{BaCO_3}}$ vs. sample depth.

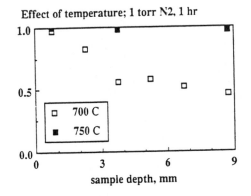

Figure 4: $\dfrac{I_{123}}{I_{123} + I_{BaCO_3}}$ vs. sample depth.

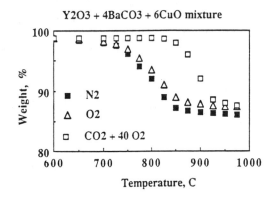

Figure 7:   TGA Results obtained with Precursor A.

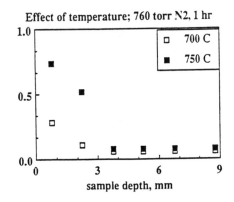

Figure 5: $\dfrac{I_{123}}{I_{123} + I_{BaCO_3}}$ vs. sample depth.

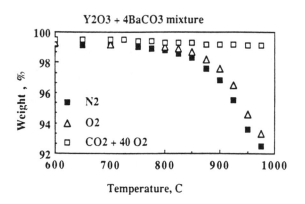

Figure 8:   TGA Results obtained with Precursor B.

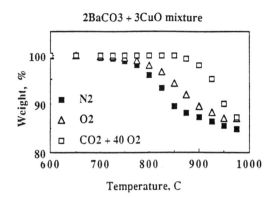

Figure 9:   TGA Results obtained with Precursor C.

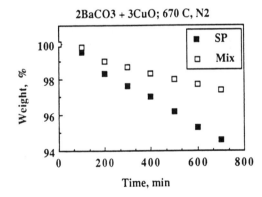

Figure 10:   Effect of Mixing on the Reaction between
             $BaCO_3$ and CuO.

SYNTHESIS AND THERMAL PROPERTIES OF STRONTIUM AND CALCIUM PEROXIDES FOR THE PREPARATION OF HIGH TEMPERATURE SUPERCONDUCTORS

Warren H. Philipp ■ National Aeronautics and Space Administration, Lewis Research Center, Cleveland, Ohio 44135 and

Patricia A. Kraft ■ Cleveland State University, Cleveland, Ohio 44115

This report describes the synthesis and discusses some thermal properties of strontium and calcium peroxides (SrO_2 and CaO_2). Our interest in these peroxides is founded on their use in the preparation of the new high-temperature superconductors. In other reports we have demonstrated several advantages in substituting BaO_2 for $BaCO_3$ in the preparation of the $YBa_2Cu_3O_7$-x bulk superconductor. The required, high-purity BaO_2 is easily made by heating BaO in dry O_2, and it is commercially available. However, SrO_2 and CaO_2, which would be used in the newer superconductors, are best synthesized by first precipitating the peroxide octahydrate from a cold, aqueous ammoniacal Sr or Ca salt solution with dilute H_2O_2. Anhydrous SrO_2 is conveniently prepared by heating the octahydrate at 115°C for 24 hr, and the less stable anhydrous CaO_2 is best synthesized by drying the octahydrate at room temperature in a dry box containing an effective desiccant such as P_2O_5. The yields of anhydrous SrO_2 and CaO_2 by this method were about 66 percent based on the amount of Sr or Ca salt used. The yield of both peroxides may be increased to about 95 percent by adding a strong base, such as NaOH, which neutralizes the ammonium salt by-product. SrO_2 is considerably more thermally stable and less susceptible to hydrolysis and CO_2 pickup than CaO_2. In addition, we give a new x-ray diffraction pattern for $CaO_2 \cdot 8H_2O$ from which we calculated the lattice parameters a = 6.212830 and c = 11.0090 on the basis of the tetragonal crystal system.

In preparing the high-temperature superconductor $YBa_2Cu_3O_7$-x, we found that substituting barium peroxide, BaO_2, for the commonly used barium carbonate, $BaCO_3$, produced a better bulk superconductor in terms of greater sample homogeneity and density and improved reproducibility ([1]). Our success in using BaO_2 as the barium source in the $YBa_2Cu_2O_7$-x formulation prompted us to investigate the synthesis and properties of strontium and calcium peroxides, SrO_2 and CaO_2, for the preparation of the newer high-temperature superconductors involving these alkaline earth oxides.

The advantages of using alkaline earth metal peroxides (BaO_2, SrO_2, and CaO_2) in place of the corresponding carbonates are briefly outlined as follows:

(1) Less rigorous reaction conditions are required in preparing superconductors because the carbonates are considerably more stable than the corresponding peroxides.

(2) There is no carbon in the starting materials; thus, it should be possible to produce a carbonate-free superconductor. Deposits of nonconducting carbonate in the grain boundaries of bulk superconductors are believed to decrease the critical current.

(3) Peroxides provide an oxidizing media that favors the formation of higher oxidation states of metals, especially copper. In the $YBa_2Cu_3O_7$-x superconductor, the oxygen stoichiometry must be greater than that required for the normal valency if superconductivity is to take place ([2]).

(4) Decomposition of peroxides liberates oxygen instead of the carbon dioxide liberated during the decomposition of carbonates. With peroxides, processing becomes simpler: a muffle furnace can be used to fire compositions containing peroxides because it is not necessary to use flowing oxygen during firing to remove liberated CO_2.

Liberated CO_2 must not remain in the sample environment because it will be taken up by the material during cooling, resulting in carbonate formation. As stated before, carbonate deposites have a detrimental effect on superconducting properties.

(5) One desirable attribute of BaO_2 is that it melts before it decomposes. This affords a liquid-solid phase reaction that is more efficient than the totally solid-solid phase reaction obtained with $BaCO_3$. The more efficient liquid-solid phase reaction with BaO_2 may explain the greater homogeneity that we observed in the bulk

$YBa_2Cu_3O_7$-x superconductor when BaO_2 was used instead of $BaCO_3$ ([1]).

(6) The anhydrous peroxides CaO_2, SrO_2, and BaO_2 are less reactive and have better storage capabilities than their respective normal oxides CaO, SrO, and BaO. The alkaline earth oxides are strong bases and, as such, tend to pickup acid gases in the atmosphere, such as CO_2, to form salts (e.g., carbonates). In addition, the oxides react vigorously with water to form hydroxides. On the other hand, the anhydrous peroxides, being less basic than the oxides, are practically inert toward CO_2 pickup under ambient conditions. They also do not appreciably react with water at room temperature to produce hydroxides.

Alkaline earth peroxides undergo three general reactions as shown. With calcium peroxide used to illustrate these reactions, they are as follows:

Decomposition:

$$CaO_2 \rightarrow CaO + \frac{1}{2}\,O_2 \qquad (1)$$

Dehydration:

$$CaO_2 \cdot 8H_2O \rightarrow CaO_2 + 8H_2O \qquad (2)$$

Hydrolysis:

$$CaO_2 \cdot 8H_2O \rightarrow Ca(OH)_2 + \frac{1}{2}\,O_2 + 7H_2O \qquad (3)$$

Table 1 gives the calculated standard enthalpy of reaction ΔHF° from the National Bureau of Standards Tables ([3]). Unfortunately, the free energy of formation ΔGF° data for the alkaline earth peroxides and their octahydrates are not available; thus ΔHF° values are used instead to explain reaction trends. For all three of the peroxides and their octahydrates, the decomposition, dehydration, and hydrolysis reactions are endothermic - unlike some peroxides, which can decompose exothermically to form the oxide, for example, hydrogen peroxide

$$H_2O_2 \rightarrow H_2O + \frac{1}{2}\,O_2 \qquad \Delta HF^\circ = -98.0\ kJ\ mol^{-1}$$

$$(4)$$

The thermodynamic data for the anhydrous peroxides indicate that they are more stable than the corresponding oxides. Thus these peroxides may be stored safely without danger of violent decomposition.

There are two general methods for synthesizing peroxides ([4]). The first involves heating the oxide in a stream of pure, CO_2-free dry oxygen. This method is favored for making peroxides that are considerably more stable than the corresponding oxide—for example, for the preparation of BaO_2. In fact, BaO_2 is commercially made by heating BaO at 500 °C in flowing, pure oxygen. The synthesis of the less stable SrO_2 requires more drastic conditions, namely heating the oxide at a temperature of 350 °C in a bomb containing high-pressure oxygen (250 atm). Finally, CaO_2, the least stable of the three, cannot be conveniently synthesized by direct combination of oxide with oxygen.

The second technique, which is more suitable for the preparation of less stable peroxides involves precipitating the insoluble peroxide from aqueous solution by adding H_2O_2 to a basic solution of the metal salt. Again using CaO_2 as the example,

$$CaCl_2 + H_2O_2 \rightarrow CaO_2\ (hydrate) + 2HCl$$

$$2HCl + 2NH_3 \rightarrow 2NH_4Cl$$

Addition of aqueous ammonia to neutralize the HCl, forces the reaction to favor the precipitation of the peroxide hydrate.

This report presents various aspects of the preparation of both the hydrated and anhydrous calcium and strontium peroxides using the second synthesis technique. Since the chemistry of CaO_2 is more complicated because it is less stable than SrO_2, it was studied in greater detail.

The thermal decomposition properties of SrO_2 and CaO_2 were also investigated. A new x-ray diffraction powder pattern was determined for $CaO_2 \cdot 8H_2O$, and lattice parameters were calculated from it. Because BaO_2 is commercially available as a pure anhydrous powder, no effort was made to investigate the synthesis and chemical properties of this compound.

EXPERIMENTAL PROCEDURE

The synthesis of strontium and calcium peroxides was accomplished by a modification of the method outlined in Brauer's Handbook of Preparative Chemistry ([4]). Calcium chloride dihydrate ($CaCl_2 \cdot 2H_2O$) or strontium nitrate ($Sr(NO_3)_2$ (0.4 mol)) was dissolved in 1.5 liters of deionized water. An excess of aqueous ammonia (ammonium hydroxide, NH_4OH), 2 mol in 600 mℓ water, was added to the stirred salt solution. No precipitated Ca or Sr salts (carbonates) were observed. In all of our cases, the solution remained clear. The solution was adjusted to the desired temperature by either heating it in a water bath or cooling it in an ice bath. Because of the ease with which the $CaO_2 \cdot 8H_2O$ loses water even under water, the octahydrate was precipitated in ice cold solution. To maintain the solution temperature at close to 0 °C, the ammoniacal $CaCl_2$ solution was kept in the freezer until some ice formed. To the well-stirred, cold solution, ice cold 3 percent H_2O_2 containing a slight excess of H_2O_2 (0.45 mol) was added at a rate of about two drops per second. The pearly white, crystalline hydrated peroxide was allowed to settle, and excess solution was removed via decantation. The crystals were filtered using a suction flask, sucked as dry as possible, washed with about 200 mℓ water containing several milliliters of NH_4OH, and then, again, sucked as dry as possible. After the peroxide was removed from the filter, it was either air dried by allowing it to stand at ambient conditions in the laboratory or dried under specified conditions (see Tables 2 and 4). The SrO_2 and the other CaO_2 preparations were synthesized in the same way except at different temperatures. For these preparations, the $Sr(NO_3)_2$ or $CaCl_2$ solution was maintained at the indicated temperature while the H_2O_2 solution was added at room temperature.

We took special care to prevent the $CaO_2 \cdot 8H_2O$ from dehydrating so that we could generate a new x-ray powder diffraction pattern of this compound. It was important that $CaO_2 \cdot 8H_2O$ remain as the pure octahydrate throughout the manipulation to ensure that the pattern was a true representation of the octahydrate. After filtration, the $CaO_2 \cdot 8H_2O$ crystals were kept moist. The moist peroxide was stored in a tightly capped plastic bottle that was kept cold in a freezer. The diffraction pattern of the wet crystals was recorded immediately after the octahydrate was removed from the freezer.

In other experiments involving $CaO_2 \cdot 8H_2O$, the moist crystals were removed from the filter and treated as indicated in the Results section. In most cases, further treatment involved heating the crystals at various temperatures in a drying oven or drying them in a flowing-nitrogen dry box containing P_2O_5 drying agent in a dish.

The peroxide yield was considerably less than stoichiometric based on the limiting reagent $CaCl_2 \cdot 2H_2O$, thus indicating some reverse reaction with the NH_4Cl by-product. We found that adding a strong base (NaOH solution) to increase pH, increased the yield to near stoichiometric. This was accomplished by the dropwise addition of 500 mℓ of a solution containing 0.75 mol of NaOH to the well-stirred precipitated peroxide suspension just following the addition of H_2O_2. Slightly less than the stoichiometric amount of NaOH was added so as to prevent the precipitation of calcium compounds other than $CaO_2 \cdot 8H_2O$, namely $Ca(OH)_2$ and $CaCO_3$. If sodium contamination presents a problem, an organic strong base such as tetramethylammonium hydroxide ($(CH_3)_4NOH$) may be substituted for NaOH.

The anhydrous CaO_2 was also directly precipitated from aqueous solution. The previously described method was employed except that it was done at higher solution temperature (e.g., 70 °C).

$SrO_2 \cdot 8H_2O$ was synthesized in a fashion similar to that for $CaO_2 \cdot 8H_2O$. However, because of its lesser tendency to dehydrate, it was unnecessary to keep the $SrO_2 \cdot 8H_2O$ moist and cold. The diffraction pattern was taken on the dried and heat treated samples and compared with the known patterns for the octahydrate and anhydrous SrO_2.

We determined the purity and degree of hydration of both peroxides by firing them for 12 hr at 1000 °C in air, thereby converting the strontium and calcium compounds to their normal oxides, SrO and CaO. X-ray diffraction was used to identify various components resulting from different runs. These components include peroxides, hydroxides, carbonates, and oxides. Particle morphology was investigated with scanning electron microscope (SEM) and optical photomicrographs. Thermogravimetric analysis (TGA) runs in helium (30 cm^3/min) at a heating rate of 10 °C/min established the decomposition temperatures of anhydrous SrO_2 and CaO_2.

RESULTS

Synthesis of SrO_2

Addition of H_2O_2 to ammoniacal $Sr(NO_3)_2$ solution (as described in the experimental section) produced pearly white, crystalline $SrO_2 \cdot 8H_2O$. The crystals settled rapidly and were readily isolated by filtration. The theoretical yield was about 66 percent based on the amount of $Sr(NO_3)_2$ starting material. Table 2 gives the effects of various treatments on the degree of hydration of $SrO_2 \cdot XH_2O$. The value of X in the table corresponds to the degree of hydration, X = moles H_2O/moles SrO_2. This value is calculated from the weight loss when $SrO_2 \cdot XH_2O$ is fired to SrO. Values of X less than 8 probably represent a mixture of $SrO_2 \cdot 8H_2O$ and anhydrous SrO_2 rather than the presence of a definite hydrated peroxide with less than eight waters of hydration.

Freshly precipitated $SrO_2 \cdot 8H_2O$ loses water of hydration gradually when it sits in a crystallizing dish at ambient laboratory conditions (about 21 °C). After 2 days the value of X drops from 8.00 to the range 7.05 to 7.70. The x-ray diffraction pattern, however, showed only the presence of the octahydrate. When freshly precipitated $SrO_2 \cdot 8H_2O$ is placed in a dry box at ambient temperature with flowing dry nitrogen and with P_2O_5 drying agent, water of crystallization is gradually lost. After 5 days the degree of hydration decreases to 0.29, and, according to x-ray diffraction, anhydrous SrO_2 becomes the major component. On drying in air at 100 °C for 24 hr, only a small amount of water (X = 0.14) re-

mains, and at 115 °C for 24 hr, pure anhydrous SrO_2 is produced. Up to this point, no decomposition products $(Sr(OH)_2$ or SrO) were observed. And, at this point, x-ray diffraction showed anhydrous SrO_2 as the sole component of the material. Chemical analysis gave 86.94 percent SrO (86.62 percent is the theoretical value), and the results of atomic adsorption analysis were also in good agreement: 73 percent Sr (whereas the theoretical value is 73.25 percent Sr). For superconductor formulations where a pure SrO_2 of known composition is required, we recommend drying the freshly precipitated hydrated SrO_2 in air at 115 °C for 24 hr. Raising the drying temperature to 125 °C resulted in a calculated negative value of X = −0.15. This denotes that in addition to complete dehydration, possibly a small amount of the normal oxide SrO was present probably because of a slight decomposition of the peroxide. TGA analysis indicated that anhydrous SrO_2 begins rapid decomposition at about 400 °C.

The theoretical yield of SrO_2 was increased from 66 percent to about 95 percent by neutralizing most (about 94 percent) of the NH_4NO_3 reaction by-product with the strong base, NaOH, thereby increasing the pH of the solution:

$$NH_4NO_3 + NaOH = NaNO_3 + NH_3 + H_2O$$

We expect an increase in pH to drive the overall SrO_2-producing reaction to the right, thereby favoring the SrO_2 yield:

$$Sr(NO_3)_2 + H_2O_2 + 2NH_3 + 8H_2O$$
$$= SrO_2 \cdot 8H_2O + 2NH_4NO_3$$

Neutralization with NaOH was not investigated further.

Synthesis of CaO_2

The relative instability of CaO_2, especially when compared with SrO_2, warranted a more extensive investigation into the chemistry of CaO_2. It is the least stable of the alkaline earth peroxides, especially with regard to hydrolysis, decomposition, and loss of water of hydration.

Table 3 lists the degree of hydration (value of X in the formula $CaO_2 \cdot XH_2O$) when the peroxide is precipitated at various solution temperatures. It is evident that the average degree of hydration of the precipitated CaO_2 depends on the solution temperature. When the precipitation is carried out at 0 °C, pearllike crystals of the octahydrate form. Like the $SrO_2 \cdot 8H_2O$ precipitate, the insoluble, hydrated CaO_2 crystals settle rapidly and are easily filtered. At this time, we attribute the unknown x-ray diffraction pattern to $CaO_2 \cdot 8H_2O$. In a later part of this report, this supposition is further substantiated. As the precipitation temperature is increased, especially above 50 °C, the beige-colored anhydrous peroxide precipitates. The insoluble anhydrous peroxide is less crystalline and is more difficult to filter. For precipitation above 50 °C, the x-ray diffraction pattern shows only the presence of the anhydrous peroxide. On the basis of the value of X, practically pure anhydrous CaO_2 is precipitated directly at the maximum temperature of our experiments, 70 °C. An important aspect to be considered in the synthesis of CaO_2 is that the peroxide yield decreases with increasing precipitation temperature; this yield is calculated from the amount of $CaCl_2 \cdot 8H_2O$ used as the limiting reagent. The yield ranges from 67.3 percent at 0 °C to 51.0 percent at 70 °C. As was the case with the deposition of $SrO_2 \cdot 8H_2O$ described earlier, increasing the pH of the solution by neutralizing 90 percent of the NH_4Cl by-product with aqueous NaOH increased the calcium peroxide yield in cold solution to about 95 percent of theoretical based on the amount of $CaCl_2 \cdot 2H_2O$ starting material. Again, the results of neutralization experiments were not studied further.

Figure 1 presents micrographs of the octahydrate and anhydrous calcium peroxide. Parts (a) and (b) show the octahedral crystal plates of freshly precipitated $CaO_2 \cdot 8H_2O$ from cold solution (0 °C). Figure 1(a) shows transparent octahedral crystals; Figure 1(b) shows the same material at higher magnification. More detail can be seen of the morphology of the eight-sided plates of $CaO_2 \cdot 8H_2O$. Parts (c) and (d) show the CaO_2 that was precipitated from hot solution (70 °C). It was sur-

prising that the CaO_2 precipitated in the form of spheres instead of well-defined crystals. Such a phenomenon suggests that when the H_2O_2 solution was added to the hot ammoniacal $CaCl_2$ solution, the octahydrate $CaO_2 \cdot 8H_2O$ formed first, then melted in its own water of crystallization with the formation of small suspended spherical droplets. As the hydrated water was lost, the liquid droplets solidified into spheres of anhydrous CaO_2. A liquid phase was observed when the octahydrate was heated to 90 °C in the oven prior to thermal dehydration to the anhydrous peroxide. It is apparent that this liquid phase is molten hydrated CaO_2.

The final investigation into calcium peroxide chemistry concerned its thermal decomposition. The results (Table 4) illustrate the ease with which hydrated CaO_2 loses water of hydration. It is difficult to obtain the pure octahydrate because attempts to remove excess absorbed water from wet, freshly prepared $CaO_2 \cdot 8H_2O$ resulted in some dehydration. The original sample used in the thermal decomposition studies was octahydrate freshly precipitated from cold solution at 0 °C and dried at room temperature overnight. The composition of this material was $CaO_2 \cdot 7.4H_2O$. To illustrate the ease with which this material loses water, we left a sample of $CaO_2 \cdot 7.4H_2O$ standing for 5 days on a lab bench. It lost most of its water of hydration, and the x-ray diffraction pattern showed only the presence of anhydrous CaO_2. The same material when placed in a flowing-nitrogen dry box containing P_2O_5 drying agent for 48 hr at room temperature lost practically all of its hydrated water (Table 4, X = 0.11). Thus, it is not surprising that thermal dehydration takes place at elevated temperatures (Table 4). Above 100 °C some decomposition (eq. (3)) takes place as indicated by $Ca(OH)_2$ showing as a minor component. At the highest drying temperature in the table, 120 °C, decomposition of the hydrated peroxide to $Ca(OH)_2$ followed by CO_2 pickup from the atmosphere becomes important. These reactions are indicated by $CaCO_3$ showing as a minor phase in the x-ray diffraction pattern. From these results, it is apparent that $CaO_2 \cdot 8H_2O$ should be dehydrated in a dry, CO_2-free atmosphere at a relatively low temperature to obtain pure, anhydrous CaO_2.

Thermal decomposition data for CaO_2 and SrO_2 are summarized in Table 5. However, these decomposition temperatures are for the pure compounds. When mixed with copper oxide (CuO), the decomposition temperatures are probably considerably lower.

The last aspect of this report concerns the new x-ray diffraction pattern attributed to calcium peroxide octahydrate, $CaO_2 \cdot 8H_2O$. We were unable to verify an exact stoichiometry for the octahydrate because of the ease with which dehydration took place during attempts to remove excess water from the freshly precipitated compound. Because of the isostructural similarity of the unknown $CaO_2 \cdot XH_2O$ with tetragonal octahydrates of other alkaline earth peroxides, $SrO_2 \cdot 8H_2O$ and $BaO_2 \cdot 8H_2O$, we attributed our new x-ray powder pattern (Table 6) to $CaO_2 \cdot 8H_2O$. The isostructuralism was confirmed through a lattice parameter program with a diffractometer extrapolation function. Photomicrographs of the CaO_2 hydrate show that its platelet morphology is similar to that described for $SrO_2 \cdot 8H_2O$ (JCPDS 25-905). This platelet morphology supports the conclusion of preferred orientation occurring and explains the well-defined, integrated-intensity results for the low index (0.0.1) planes. As expected, the lattice parameters of $CaO_2 \cdot 8H_2O$ would be smaller than those for $SrO_2 \cdot 8H_2O$ and $BaO_2 \cdot 8H_2O$ because the Ca atom is smaller than the Sr and Ba atoms.

CONCLUSIONS

The most important practical aspect of our experimental results is the development of a simple synthesis for pure CaO_2 and SrO_2. The peroxides are of particular interest as replacements for the corresponding alkaline earth carbonates for the synthesis of the new high-temperature superconductors. Their use in BiCaSrCuO and the most recent TlBaCaCuO systems should offer the same advantages as we found when BaO_2 was substituted for $BaCO_3$ in the preparation of the $YBa_2Cu_3O_{7-x}$ superconductor. Furthermore, CaO_2 and SrO_2 may be generally useful when an oxidizing alkaline fusion media is required to

synthesize compounds other than superconductors.

In view of our experimental results, we recommend the following procedure for the practical synthesis of pure SrO_2 and CaO_2. For both peroxides, the first step involves precipitation of the octahydrate by dropwise addition of H_2O_2 to a cold solution (about 0 °C) of ammoniacal $Sr(NO_3)_2$ or $CaCl_2$. The best way to prepare anhydrous SrO_2 is to dry the freshly prepared octahydrate for 24 hr at 115 °C, preferably in a CO_2-free atmosphere. In the case of the less stable CaO_2, the freshly precipitated octahydrate is best dehydrated at room temperature for at least 24 hr in a CO_2-free drybox containing an efficient desiccant such as P_2O_5. The octahydrates of both peroxides are crystalline precipitates that tend to settle out rapidly, thereby allowing most of the mother liquor to be removed by decantation. The crystalline octahydrates are easy to recover by filtration and have little tendency to peptize during filtration and washing. If peroxide yield is important, it can be increased by neutralizing most of the ammonium salt byproduct with a strong base such as NaOH. If sodium contamination is to be avoided, a strong organic base such as tetramethylammonium hydroxide $(CH_3)_4NOH$ may be used instead.

REFERENCES

1. Hepp, A.F, et al.: Advantages of Barium Peroxide in the Powder Synthesis of Perovskite Superconductors. High Temperature Superconductors, MRS Symp.Proc. vol. 99, M.B. Brodsky, et al., eds., Materials Research Society, 1988, pp. 615-618.

2. Brodsky, M., et al., eds.: High Temperature Superconductors. MRS Symp. Proc. vol. 99, Materials Research Society, 1988.

3. The NBS Tables of Chemical and Thermodynamic Properties American Chemical Society and American Institute of Physics for the National Bureau of Standards. J. Phys. Chem. Ref. Data, vol. 11, 1982, Suppl. 2.

4. Brauer, G.: Handbook of Preparative Inorganic Chemistry, Vol. 1, 2nd ed., Academic Press, New York, 1963, pp. 936-938.

5. Weast, R.C., ed.: Handbook of Chemistry and Physics. 64th ed. CRC Press Inc., Boca Raton, FL, 1983-1984.

TABLE 1. - CALCULATED STANDARD ENTHALPY OF REACTION ($\Delta HF°$)
FOR DECOMPOSITION, DEHYDRATION, AND HYDROLYSIS OF THE
ALKALINE EARTH PEROXIDES

[From National Bureau of Standards Tables ([3]).]

Reaction	Equation	$\Delta HF°$, kJ·mol^{-1}		
		CaO_2	SrO_2	BaO_2
Decomposition	(1)	17.6	41.5	80.8
Dehydration	(2)	65.6	103.3	85.7
Hydrolysis	(3)	18.0	63.6	61.1

TABLE 2. - EFFECT OF VARIOUS TREATMENTS ON SrO_2·XH_2O

Treatment	$\dfrac{\text{moles } H_2O}{\text{moles } SrO_2}$, X	Composition (from x-ray diffraction analysis)
Air dried, room temperature, 2 days	7.05 to 7.70	SrO_2·$8H_2O$
Dried in dry box over P_2O_5, room temperature, 5 days	0.29	Major: SrO_2 Minor: SrO_2·$8H_2O$
Air dried, 100 °C, 24 hr	0.14	SrO_2
Air dried, 115 °C, 24 hr[a]	0	SrO_2
Air dried, 125 °C, 24 hr	-0.15	SrO_2

[a]X-ray diffraction analysis: SrO_2 is the only phase.
Chemical analysis: 86.94 percent SrO (86.62 percent theoretical).
Atomic adsorption anlaysis: 73 percent Sr (73.25 percent theoretical).

TABLE 3. - SYNTHESIS OF CaO_2·XH_2O

Precipitation temperature, °C	$\dfrac{\text{moles } H_2O}{\text{moles } CaO_2}$, X	Composition (from x-ray diffraction analysis)	Yield, percent[a]
0	7.4	(b)	67.32
5	6.3	(b)	-----
10	5.1	(b)	64.23
50	.42	CaO_2	-----
60	.13	CaO_2	-----
65	.10	CaO_2	54.09
70	.09	CaO_2	51.02

[a]Based on amount of $CaCl$·$2H_2O$ (limiting reagent) used.
[b]Attributed to CaO_2·$8H_2O$.

TABLE 4. - EFFECT OF HEAT TREATMENT ON MOLES H_2O/MOLES CaO_2

[Starting material: 7.4 mol H_2O/mol CaO_2. Hydrated CaO_2 tends to dehydrate; reacts with ambient CO_2 to form $CaCO_3$.]

Temperature, °C	Duration, hr	$\dfrac{\text{moles } H_2O}{\text{moles } CaO_2}$, X	Composition (from x-ray diffraction analysis)
90	24	0.15	CaO_2
100		.13	Major: CaO_2 Minor: $Ca(OH)_2$
107		.098	Major: CaO_2 Minor: $Ca(OH)_2$
120		.18	Major: CaO_2 Minor: $CaCO_3$
Room temperature	48	.11	CaO_2

TABLE 5. - DECOMPOSITION TEMPERATURES TO FORM OXIDES
([5])

$2MO_2 \rightarrow 2MO + O_2$		$2MCO_3 \rightarrow MO + CO_2$	
Compound	Decomposition temperature, °C	Compound	Decomposition temperature, °C
BaO_2	[a]800	$BaCO_3$	1450
SrO_2	[b]400	$SrCO_3$	1340
CaO_2	[b]330	$CaCO_3$	899

[a]Melting point, 450 °C
[b]Determined from our own TGA data.

TABLE 6. - X-RAY POWDER DATA FOR
CaO_2·$8H_2O$

[Crystal system; tetragonal lattice parameters (Å); a = 6.212830±7.628441×10^{-3}; c = 11.00990±1.083599×10^{-2}; all peaks attributed to CaO_2·$8H_2O$; no known impurity peaks for CaO_2, $CaCO_3$, and $Ca(OH)_2$ were observed.]

2-theta (obs)	d(Å)	hkl	I/I_0
14.287	6.1993	100	1
16.182	5.4774	002	100
20.295	4.3758	110	1
21.656	4.1037	102	2
26.003	3.4268	112	13
28.759	3.1043	200	1
32.601	2.7467	004	63
33.188	2.6994	211	11
35.724	2.5134	104	7
36.250	2.4782	212	1
38.695	2.3270	114	23
40.765	2.2135	213	11
41.118	2.1953	220	1
44.017	2.0572	204	11
44.407	2.0400	222	2
46.182	1.9657	310	5
46.496	1.9532	214	10
47.003	1.9333	302	1
49.248	1.8503	312	3
52.029	1.7577	106	18
53.111	1.7244	215	5
54.232	1.6914	116	2
55.532	1.6549	304	1
58.495	1.5779	206	1
60.502	1.5290	216	8
60.715	1.5279	216	4
62.120	1.4942	410	1
63.528	1.4645	324	<1
66.032	1.4149	332	<1
67.323	1.3909	420	1
68.215	1.3737	306	6
68.602	1.3669	217	6
68.831	1.3663	217	3
69.467	1.3531	404	2
70.083	1.3416	108	10
70.333	1.3407	108	5
71.328	1.3223	118	<1
73.186	1.2922	334	1
75.708	1.2563	326	1
77.422	1.2317	218	2
77.750	1.2304	218	1

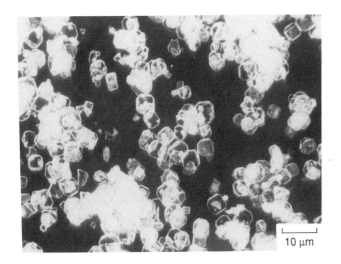

(a) CaO$_2$ · 8H$_2$O precipitated from O °C solution.

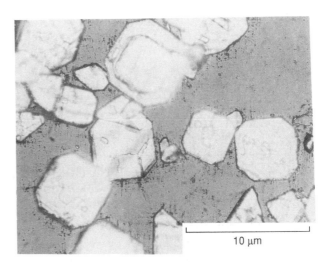

(b) **Enlarged view of CaO$_2$ · 8H$_2$O precipitated from O °C solution.**

(c) Anhydrous CaO$_2$ precipitated from 70 °C solution.

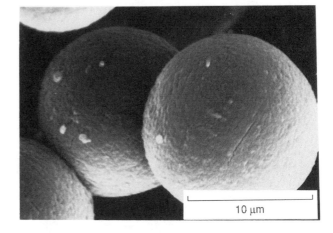

(d) Enlarged view of CaO$_2$ precipitated from 70 °C solution.

Figure 1.—Optical photomicrographs of precipitated anhydrous and octahydrate calcium peroxide.

PHASE RELATIONSHIPS IN THE Ba-Y-Cu-O SYSTEM IN AIR

Ahmed M. Gadalla†, Paisan Kongkachuichay, and Turi Hegg ■ Department of Chemical Engineering, Texas A&M University, College Station, TX 77843

Compatible phases as well as melting relationships in the system $BaO-Y_2O_3-Cu-O$ in air were studied as a function of temperature by means of simultaneous thermal analysis and powder X-ray diffraction. $Ba_3YCu_2O_x$ (312), and $Ba_2YCu_3O_x$ (213), BaY_2CuO_5 (121) were found to melt peritectically at 1096, 1002 and 1265°C respectively. The following tie lines were established: 213 - $BaCuO_2$, 213 - CuO, 213 - 121, 121 - CuO, 121 - $Y_2Cu_2O_5$, 121 - Y_2O_3, 121 - BaY_2O_4, 121 - 312, 121 - $BaCuO_2$, 312 - BaY_2O_4, 312 - $Ba_2Y_2O_5$, 312 - $Ba_4Y_2O_7$, 312 - BaO, and 312 - $BaCuO_2$. The phase relationships in this system were projected on the ternary diagram BaO - $1/2Y_2O_3$ - CuO. The primary phase fields, invariant points and isotherms on liquidus surface were determined.

1. Introduction

Since the discovery of the high temperature superconductor $Ba_2YCu_3O_x$, phase relationships in the system Ba-Y-Cu-O were studied by many groups. Accurate phase diagrams provide necessary information required for synthesis and processing. The compatible phases and their limits of stability depend on not only composition and temperature but also oxygen partial pressure. To study the phase relationships one can keep the temperature constant and study the phase changes as a function of pressure (isothermal section) as done by Ahn et al.[1] or to keep the pressure constant and study the phase changes as a function of temperature (isobaric section) as done in the present and most of previous studies[2-15]

Roth et al.[2, 3] constructed a tentative diagram and specified some primary phase fields. They showed $Ba_2YCu_3O_x$ (213) to melt peritectically at 1002°C to liquid and BaY_2CuO_5 (121) which melts also peritectically to Y_2O_3 plus liquid at about 1275°C. They also concluded that $Ba_3YCu_2O_x$ is chemically and structurally an oxycarbonate. The same kind of work was done in the Cu-rich portion by Aselage et al.[4] but there is no agreement for melting relationships. Frase et al.[5,6] and Wang et al.[7] published subsolidus relationships in this system. The compounds Ba_2CuO_3 and Ba_3CuO_4 were reported by Frase et al.[5,6] to decompose above 850°C, Wang et al.[7] reported that Ba_2CuO_3 could not be produced without Li_2CO_3 as mineralizer. Arjomand and Machin[8] reported that $BaCuO_2$ picked up oxygen at 350°C and lost it at about 600°C (presumably in air). They reported that at high oxygen pressure $BaCuO_{2.5}$ can be

formed. While Frase et al.[5,6] and Roth et al.[2,3] agree on the existence of $Ba_4Y_2O_7$, Roth et al.[2,3] and Wang et al.[6] agree on the existence of $Ba_2Y_2O_5$ and only Roth et al. showed $Ba_3Y_4O_9$ to exist. Recently, Lay and Renlund[9] determined the liquidus temperatures near the composition $Ba_2YCu_3O_x$ for oxygen partial pressures from 2 x 10^{-4} to 1 bar, and some parts of their results (at P_{O_2} = 0.21 bar) will be also compared with this study.

Arjomand and Machin[8] reported the existence of $Y_2Cu_2O_5$ in air at 850°C and stated that under 400 atm of oxygen at 800°C $YCuO_3$ can be reached. Roth et al.[2] showed $Y_2Cu_2O_5$ to melt peritectically at about 1135°C compared to 1122°C reported by Aselage et al.[4]. Gadalla and Kongkachuichay[10] proved that $Y_2Cu_2O_5$ is stoichiometric and melts in air peritectically at about 1110°C. They found also that it reacts with excess CuO (reversibly) forming a compound having composition $YCu_2O_{2.5}$ which exists between 990 and 1105°C in air.

Reviewing the subsolidus results by previous reports[2,3,5-7] indicates conflicts regarding the existence of compatible phases which need to be clarified by this work as will be explained.

2. Experimental Techniques

All existing binary and ternary compounds in this system were prepared by mixing $BaCO_3$ with Y_2O_3 and CuO at corresponding stoichiometric ratios. These mixtures were fired in air, at temperature lower than that required for liquid formation, with intermediate grinding until a constant weight was achieved. They were examined by X-ray diffraction to confirm that the reactions were complete.

†Author to whom correspondence should be addressed

To resolve the conflicts about the existing tie lines, mixtures of phases which might be compatible were mixed and fired at a temperature below that required for liquid formation. Then they were examined by X-ray diffraction to determine if the constituents remain indefinitely (compatible) or a reaction occurs to give other phases. Accordingly, the tie lines were established.

More than 105 mixtures consisting of the corresponding binary compounds, ternary compounds, and/or the starting materials were thermally analyzed using DTA and TG. Some mixtures were examined using X-ray diffraction and the system BaO-Y_2O_3-CuO was then constructed. The diagrams show the composition and phase changes that occur in any mixture when it is heated or cooled. Since the principles involved with such non-condensed systems have been discussed previously[10-13] a detailed discussion is not attempted here. The following generalizations, however, may be stated for a quaternary system:

1. When five condensed phases coexist with the gas phase, the system will be invariant and there will be only one temperature and oxygen pressure at which these six phases can exist. Examples are a quaternary eutectic point and a quaternary reaction point. Any slight change in temperature or oxygen pressure will cause this point to be missed. This situation did not exist in the isobar under consideration (in stagnant air).

2. When four condensed phases coexist with the gas phase, the system will be monovariant, and at any given oxygen pressure these five phases will coexist at one temperature only. The four condensed phases form a conjugation tetrahedron. If we start with three solids lying on the face BaO-Y_2O_3-CuO, the first liquid to appear will lie inside the quaternary system. Accordingly several conjugation tehedra will form and each tetrahedron join the liquid composition with three compatible solid phases existing on the face BaO-Y_2O_3-CuO. The three solids exist at the apices of a triangle bounded by three tie lines. The composition of the liquid phase projected on the face BaO-Y_2O_3-CuO (Figure 1) appears as the point of intersection of three boundaries. Under equilibrium conditions an isothermal oxygen loss occurs giving a nearly vertical step on the thermogravimetric curves. The differential thermal analysis curve indicates if this situation corresponds to endothermic or exothermic change. In the present study the onset peak temperatures obtained from the thermal curves were recorded as the monovariant transitions.

3. When three condensed phases coexist with the gas phase, the system is bivariant, and at any given oxygen pressure these four phases will coexist generally over a wide temperature range. At each temperature the condensed phases form a conjugation triangle. A progressive change in weight occurs as the temperature increases giving wide or diffused DTA peaks. This situation occurred in the present system for mixtures having composition deviating from those corresponding to liquid compositions existing in monovariant situations (discussed above). For such mixtures, at the monovariant temperature initial isothermal melting occurs giving two residual solids with the liquid phase. Its composition moves inside the quaternary system. Its projection on the face BaO-Y_2O_3-CuO is shown in Figure 1 as a boundary curve. Three boundary curves meet at a ternary monovariant point.

4. When two condensed phases coexist with the gas phase, the system is trivariant. At constant oxygen pressure the three phases will coexist over a range of temperatures. At each temperature a tie line joining the two condensed phases exists. This situation will occur if two solids or a solid and a liquid exist in equilibrium and can be seen in the sections shown in Figures 2, 3 and 4.

5. When one condensed phase coexist with the gas phase, the system is tetravariant, and at constant oxygen pressure the two phases will coexist over a range of temperatures as in the previous case.

In this study the thermal analyses were performed by using the Netzsch Simultaneous Thermal Analyzer, STA 409. The sample sizes ranged from 100 to 200 mg with heating and/or cooling rates of 3 K/min to obtain DTA and TG simultaneously. The reference material was calcined kaolin, and the crucibles for holding samples were made from high purity 99.5% alumina supplied by Netzsch. The operating atmosphere was stagnant air. To produce more accurate temperatures which are closer to equilibrium values, a low heating rate was selected and instead of reporting the peak temperature, the temperarure at which any reaction begins was reported. These temperatures can be easily determined from the first derivative.

3. Results and Discussion

All the binary and ternary compounds, which were reported to exist in this quaternary system, were prepared at 800 to 950°C (below their melting points). Powder X-ray diffraction was performed to confirm the existing phases. The ternary compounds BaY_2CuO_5 (121), $Ba_2YCu_3O_x$ (213),and $Ba_3YCu_2O_x$ (312) were successfully prepared at 950°C and after soaking in air at 400°C for 60 h, the 213 and 312 compounds picked up oxygen to the approximate compositions $Ba_2YCu_3O_{6.9}$ and $Ba_2YCu_2O_7$ respectively but the 121 compound did not change its weight.

Since the present results indicated the absence of $Ba_3Y_4O_9$, Ba_2CuO_3 and Ba_3CuO_4 in air, the tie lines $Ba_4Y_2O_7$-Ba_3CuO_4, $Ba_4Y_2O_7$-Ba_2CuO_3, and $Ba_3Y_4O_9$-312 do not exist. Mixtures of $Ba_2Y_2O_5$-121 and BaY_2O_4-312 (containing the composition corresponding to the intersection of the lines joining $Ba_2Y_2O_5$-121 and BaY_2O_4-312) were mixed and fired at 900°C for 50 h with intermediate grinding. The

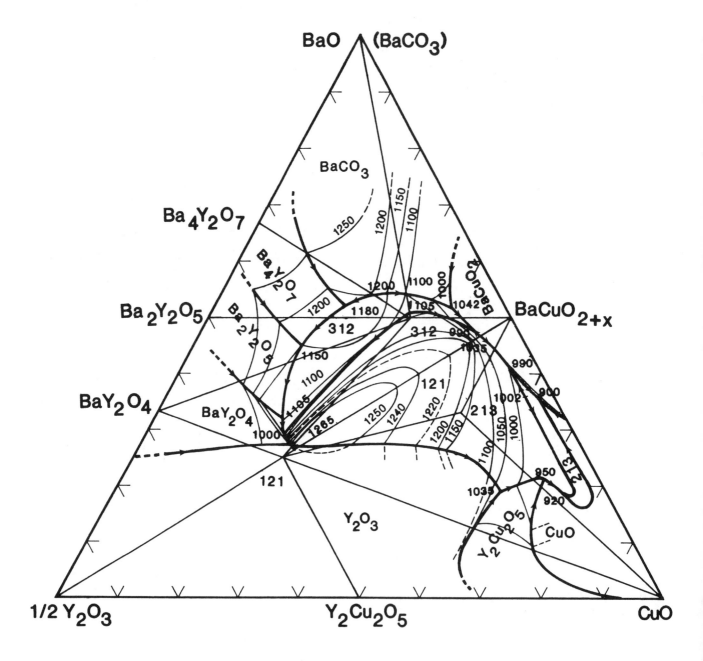

Figure 1. The pseudo ternary diagram of the BaO (BaCO₃)-Y₂O₃-CuO system constructed by projecting the results obtained by heating fully oxidized mixtures in air.
213: $Ba_2YCu_3O_x$ 121: BaY_2CuO_5
312: $Ba_3YCu_2O_x$

green color changed gradually to black and the X-ray diffraction patterns showed the existence of only BaY_2O_4 and 312 in both mixtures. The results indicated that BaY_2CuO_5 reacted with $Ba_2Y_2O_5$ forming $Ba_3YCu_2O_x$ and BaY_2O_4. Accordingly, BaY_2O_4 and 312 are compatible and the line joining them is a true tie line. This result is in agreement with Roth et al.[2-3] and Frase et al.[5-6] but not with Wang et al.[7]. It seems that the conclusion by Wang et al.[7] that $Ba_2Y_2O_5$-121 join exists, was based on observing the green phase microscopically in their mixtures. It should be noted, however, that traces of this phase will be present if the reaction is not complete and their presence indicates that equilibrium conditions were not achieved. Using the above technique, the following tie lines were also confirmed: $Ba_2Y_2O_5$-312, BaO ($BaCO_3$)-312 and $BaCuO_2$-121. All existing tie lines (those previously established and those established by the present study) are shown in Figure 1. It is to be noted that in presence of dry air, BaO transform slowly to $BaCO_3$ and accordingly $BaCO_3$ is added between brackets wherever BaO was mentioned and is also added in Figure 1. The other phases reported in the introduction to exist as oxycarbonates were shown here as oxides and the present work concentrates only on the compatible phases in air without studying the CO_2 content and the extend of solid solutions which were covered by previous workers.

In order to complete the whole ternary phase diagram of the BaO-Y_2O_3-Cu-O system, the binary and ternary compounds were used to prepare mixtures of compatible phases. Thermal analyses were performed to establish their thermal effects in air and monovariant situations (case 2, see above) were obtained and used in constructing the phase diagram. The results were projected on the ternary diagram shown in Figure 1. The primary phase fields and boundary lines, cooling direction along the boundary lines, invariant points and isotherms on the liquidus surface are shown. Referring to results by Roth et al.[2,3], the 121 primary phase field was found to extend to the lower part of the diagram, the $Y_2Cu_2O_5$ primary phase was shifted to be near the CuO corner. The primary phase fields of BaY_2O_4, $Ba_2Y_2O_5$, $Ba_4Y_2O_7$, and 312 were added on the diagram.

In the CuO-$BaCuO_2$-213 triangle four monovariant points are included; one ternary eutectic point at 900°C and three ternary reaction points at 920, 950, and 990°C. A saddle point (at 1002°C) exists along the line separating the primary fields of 121 and 213. The ternary eutectic point is the lowest melting point at which the CuO, $BaCuO_2$, and 213 are in equilibrium with a liquid. At the first ternary peritectic point (920°C) CuO, 121, 213, and liquid coexist in equilibrium in air. At this point the following reaction occurs:

$$213 + CuO \quad \underset{cooling}{\overset{heating}{\rightleftarrows}} \quad 121 + liquid.$$

Similarly, at 950°C, CuO, $Y_2Cu_2O_5$, 121, and liquid coexist in equilibrium with oxygen partial pressure of 0.21. They represent the reaction

$$121 + CuO \quad \underset{cooling}{\overset{heating}{\rightleftarrows}} \quad Y_2Cu_2O_5 + liquid.$$

At 990°C $BaCuO_2$, 121, 213, and liquid coexist in equilibrium with this pressure. The corresponding reaction is

$$213 + BaCuO_2 \quad \underset{cooling}{\overset{heating}{\rightleftarrows}} \quad 121 + liquid.$$

In this triangle the transition temperatures are generally in good agreement with Roth et al.[1,2] but they are lower than those of Aselage et al.[4] and Lay et al.[9]. However, both Aselage et al. and Lay et al. used heating rates of 10 °C/min which tends to shift the transitions to higher temperatures (which reached in some cases 35°C).

A ternary point corresponding to the existence of $Y_2Cu_2O_5$, 121, Y_2O_3, and liquid is located at 1035°C inside the CuO-213-121 triangle. The corresponding reaction is

$$121 + Y_2Cu_2O_5 \quad \underset{cooling}{\overset{heating}{\rightleftarrows}} \quad Y_2O_3 + liquid$$

On the line 121-$BaCuO_2$ two invariant points exist at 1035 and 1265°C. They correspond to two saddle points: one on the trough between $BaCuO_2$ and 121, and the second on the trough between 121 and Y_2O_3.

In the 121-312-BaY_2O_4 triangle, one ternary eutectic point and two ternary reaction points are found at 1000, 1105, and 1240°C respectively. At 1240°C, Y_2O_3, BaY_2O_4, 121 and liquid are in equilibrium in air, corresponding to the reaction

$$BaY_2O_4 + 121 \quad \underset{cooling}{\overset{heating}{\rightleftarrows}} \quad Y_2O_3 + liquid$$

This reaction was reported by Roth et al.[2] to occur also at about 1240°C. At 1000, 1105°C, the following reactions occurred respectively:

$$BaY_2O_4 + 121 + 312 \underset{\text{cooling}}{\overset{\text{heating}}{\rightleftarrows}} \text{liquid,}$$

$$312 + Ba_2Y_2O_5 \underset{\text{cooling}}{\overset{\text{heating}}{\rightleftarrows}} BaY_2O_4 + \text{liquid}$$

Mixtures on the line 121-312 showed the existence of two transitions at 1105 and 990°C. They correspond to a saddle point on the trough between 312 and 121, and an eutectic point corresponding to the existence of $BaCuO_2$, 121, 312 and liquid respectively. If the liquid at 1105°C cools down to the right-hand side of the 1105°C point until it reaches temperature of 990°C, $BaCuO_2$ will crystallize. On the other hand, BaY_2O_4 will crystallize at 1000°C, which is the ternary eutectic point at the other end of the boundary line.

On the outside shell of the 312 primary phase field, another two ternary peritectic points and one ternary eutectic point are detected. The corresponding reactions are

$$312 + BaCuO_2 + BaCO_3 \underset{\text{cooling}}{\overset{\text{heating}}{\rightleftarrows}} \text{liquid} \quad \text{(at 1000°C),}$$

$$312 + Ba_2Y_2O_5 \underset{\text{cooling}}{\overset{\text{heating}}{\rightleftarrows}} Ba_4Y_2O_7 + \text{liquid} \quad \text{(at 1150°C)}$$

$$312 + Ba_4Y_2O_7 \underset{\text{cooling}}{\overset{\text{heating}}{\rightleftarrows}} BaCO_3 + \text{liquid} \quad \text{(at 1180°C).}$$

Accordingly, the behavior of the ternary compounds can be explained. The 121 compound was found to be stoichiometric, while the 213 and 312 compounds were non-stoichiometric. All three compounds were found to melt peritectically at 1265, 1002 and 1096°C respectively according to the following reactions:

$$121 \underset{\text{cooling}}{\overset{\text{heating}}{\rightleftarrows}} Y_2O_3 + \text{liquid,}$$

$$213 \underset{\text{cooling}}{\overset{\text{heating}}{\rightleftarrows}} 121 + \text{liquid,}$$

$$312 \underset{\text{cooling}}{\overset{\text{heating}}{\rightleftarrows}} 121 + \text{liquid.}$$

According to the diagram, the 213 compound should give a peak at 1002°C for its peritectic melting followed by complete melting at 1195°C. Extra peaks were always present at 900, 950, and/or 990°C and are due to presence of traces of unreacted intermediate compounds. 990°C represents the invariant point corresponding to 213, 121 and $BaCuO_2$, while 900°C corresponds to that of 213, $BaCuO_2$ and CuO, and 950°C corresponds to that between 213, 121 and CuO. The reported complete melting of the 213 at 1195°C is inconsistent with Lay and Renlund[9], Aselage et al.[4] and Murakami et al.[14,15] who observed Y_2O_3 and liquid phase at 1350°C, 1400°C and 1450°C respectively. According to them the primary phase is Y_2O_3 and not 121. It has to be noted that they equilibrated the corrosive liquid for long periods at high temperatures and it was documented by the present and previous authors[11,14,15] that platinum and alumina can extract copper and form solid solution and aluminates respectively. Accordingly, at these temperatures, copper oxide is extracted leaving a mixture poorer in copper oxide and its composition moves towards Y_2O_3 corner. According to the present results these new mixtures will yield Y_2O_3 as the primary phase as observed by these three research groups.

Finally, three sections are constructed to show phase changes occurring in mixtures consisting initially of 121-213 (Figure 2), $Y_2Cu_2O_5$-$BaCuO_2$ (Figure 3) and $Ba_2Y_2O_5$-CuO (Figure 4). Vertical lines represent the intersection of the section with tie lines while horizontal lines represent monovariant situations (four condensed phases). The sub-liquidus curves bound compositions containing three condensed phases and liquidus curves show the temperatures above which complete melting occurs. The latter points determine the isothermos on the liquidus surface shown in Figure 1.

4.   Conclusions

The ternary compound 121 was found to be stoichiometric, while the 312 and the 213 compounds were non-stoichiometric reaching a maximum oxygen content of $Ba_3YCu_2O_7$ and $Ba_2YCu_3O_{6.9}$ in air at 400°C. All three compounds were found to melt peritectically at 1265, 1096 and 1002°C respectively.

The following tie lines were confirmed: 213-$BaCuO_2$, 213-CuO, 213-121, 121-CuO, 121-Y_2O_3, 121-BaY_2O_4, 121-312, 121-$BaCuO_2$, 312-BaY_2O_4, 312-$Ba_2Y_2O_5$, 312-$Ba_4Y_2O_7$, 312-BaO ($BaCO_3$), and 312-$BaCuO_2$. On the other hand, the previously reported tie lines $Ba_4Y_2O_7$-Ba_3CuO_4[4-5], $Ba_4Y_2O_7$-

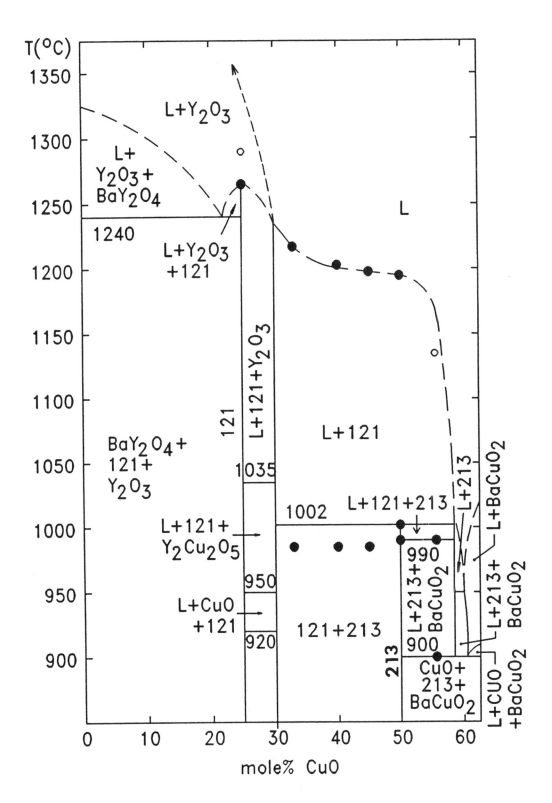

Figure 2. Section showing the behavior of mixtures containing initially BaY_2CuO_5–$Ba_2YCu_3O_x$
($\bullet$ phase change, $\circ$ partial melting)
213: $Ba_2YCu_3O_x$ 121: BaY_2CuO_5
312: $Ba_3YCu_2O_x$

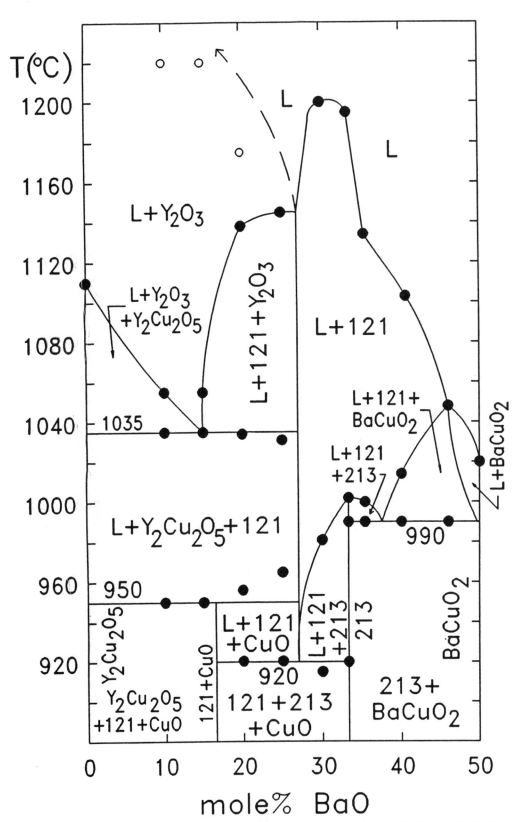

Figure 3. Section showing the behavior of mixtures containing intially $Y_2Cu_2O_5$-$BaCuO_2$.
(● phase change, ○ partial melting)
213: $Ba_2YCu_3O_x$ 121: BaY_2CuO_5
312: $Ba_3YCu_2O_x$

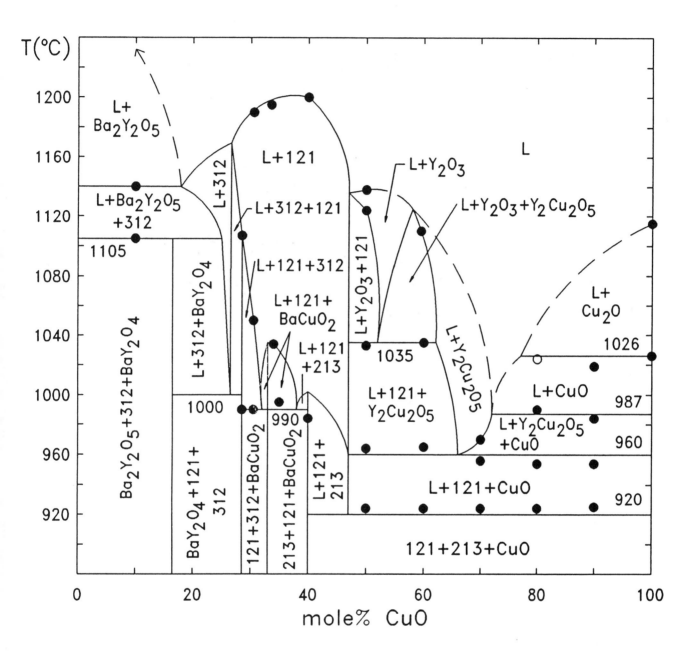

Figure 4.   Section showing the behavior of mixtures containing initially $Ba_2Y_2O_5$-CuO.
(● phase change, ○ partial melting)
213: $Ba_2YCu_3O_x$ 121: BaY_2CuO_5
312: $Ba_3YCu_2O_x$

Ba_2CuO_3[4-5], $Ba_3Y_4O_9$-312[2] and $Ba_2Y_2O_5$-121[7] were not observed.

All established tie lines, primary phase fields, boundary lines, invariant points, and isotherms on liquidus surface were projected on the quasi-ternary diagram BaO-1/2Y_2O_3-CuO. The sections of 121-213, $Y_2Cu_2O_5$-$BaCuO_2$ and $Ba_2Y_2O_5$-CuO were constructed.

Acknowledgement

We would like to thank the National Science Foundation, Grant #8717554 and the Board of Regents, Texas A&M University (AUF - sponsored Materials Science and Engineering Program).

References:

[1] B. T. Ahn, V. Y. Lee and R. Beyers, Physica C. **167**, 529 (1990).

[2] R. S. Roth, K. L. Davis and J. R. Dennis, Adv. Ceram. Mater. **2**. 303 (1987).

[3] R. S. Roth, C. J. Rawn, F. Beech, J. D. Whitler and J. O. Anderson, *Ceramic Superconductor II*, Edited by M. F. Yan (Am. Ceram. Soc., Westerville, OH, 1988) pp. 13-26.

[4] T. Aselage and K. Keefer, J. Mater. Res. **3**, 1279 (1988).

[5] K. G. Frase, E. G. Liniger and D. R. Clarke, J. Am. Ceram. Soc. **70**, C204 (1987).

[6] K. G. Frase and D. R. Clarke, Adv. Ceram. Mater. **2**, 295 (1987).

[7] G. Wang, S. J. Hwu, S. N. Song, J. B. Ketterson, L. D. Marks, K. R. Poeppelmeier and T. O. Mason, Adv. Ceram. Mater. **2**, 313 (1987).

[8] M. Arjomand and D. J. Machin, J. Chem. Soc. Dalton. **11**, 1061 (1975).

[9] K. W. Lay and G. M. Renlund, J. Am. Ceram. Soc. **73**, 1208 (1990).

[10] A. M. Gadalla and P. Kongkachuichay, J. Mater. Res., **6**, 450 (1991).

[11] A. M. Gadalla and J. White, Trans. Brit. Ceram. Soc. **65**, 1 (1966).

[12] A. M. Gadalla and N. A. Mansour, Nuclear Science and Engineering **86**, 247 (1984).

[13] A. M. Gadalla and N. A. Mansour, Ind. and Eng. Chem. Fund. **23**, 440 (1984).

[14] M. Murakami, M. Morita, K. Doi, and K. Miyamoto, Jpn. J. Appl. Phys. **28**, 1189 (1989).

[15] M. Murakami, M. Morita, and N. Koyama, Jpn. J. Appl. Phys. **28**, L1125 (1989).

XRD CHARACTERIZATION OF FLUORINE-DOPED $Y_1Ba_2Cu_3O_{7-x}$

Sendjaja Kao, Sangho Lee, and K.Y. Simon Ng ■ Department of Chemical Engineering, Wayne State University, 5050 Anthony Wayne Drive, Detroit, Michigan 48202

Evidence of incorporation of fluorine into $Y_1Ba_2Cu_3O_{7-x}$ by solid state reaction was observed using XRD. Some of the BaF_2 precursor was found to be decomposed and reacted to form a fluorine-doped 123 phase, and an amorphous-tetragonal fluorinated phase in $Y_1Ba_2Cu_3F_2O_y$ and $Y_1Ba_2Cu_3F_4O_y$ samples respectively. The different thermal behaviors of the 123 and fluorinated samples were attributed to the different phases observed.

The discovery of high-Tc superconductivity in La-Ba-Cu-O by Bernorz and Muller [1] in 1986 has created tremendous interest in superconductivity research. Within several months, the Tc was increased to above liquid nitrogen temperature (T = 77 K) with the discovery of Y-Ba-Cu-O [2] superconducting oxides, noted as either 123 or YBCO, with Tc about 92 K. Since then, many different superconducitng oxides with Tc greater than 35 K were discovered independently. Currently, thallium- and bismuth-based systems are known to contain superconducting phases with Tc greater than 100 K. There have also been several reports of superconductivity greater than 150 K, and even as high as over 250 K [3-8].

Ovshinsky et al. [7] reported a significant increase in Tc by introducing fluorine into the YBCO samples. They observed a complete loss of electrical resistance in the fluorinated sample at ca. 148 K to 168 K. Chen et al. [3] observed a superconducting anomaly at a temperature greater than 250 K when they measured the electrical properties of their $Y_5Ba_6Cu_{11}O_y$ samples in an oxygen-filled environment. However, most of these Tc enhancements cannot be easily reproduced.

In the fluorine-doped YBCO sample, mixed results have been reported regarding the incorporation of fluorine into the 1-2-3 superconducting phase by conventional solid state reaction [7-13], and regarding the effect of Tc with fluorine substitution [7,8,11-16]. However, it is agreed that small amounts of fluorine can be incorporated by annealing the YBCO sample in a fluorine-containing environment. Several groups [10,17,18] reported unsuccessful attempts to prepare fluorinated samples by solid state reaction. Hou et al. [10], based on their $19F^-$ NMR and XRD data, observed no fluorine-containing 123 phase but a high concentration of BaF_2, and significant amounts of unreacted CuO and $Y_2Cu_2O_5$ sample in their $Y_1Ba_2Cu_3F_4O_y$. However, they observed a weak intensity of X-ray peaks which is consistent with the strongest 1-2-3 pervoskite peaks. In contrast, several groups [12-14] observed an increase of lattice parameters with increasing fluorine content. Nakayama et al. [12] attributed the lattice expansion to fluorine occupying the vacant site of the YBCO lattice. In our previous study [19], we observed anomalous thermal behavior which suggests the presence of fluorine-containing 123 phase.

In this communication, we continue our study of the fluorinated phase by characterizing the samples using XRD. The results indicate the existence of different fluorine-containing 123 phases after solid state reaction.

EXPERIMENTAL

Three samples with $Y_1Ba_2Cu_3O_{7-x}$ (123), $Y_1Ba_2Cu_3F_2O_y$ (123F2), and $Y_1Ba_2Cu_3F_4O_y$ (123F4) nominal compositions were prepared by solid state reaction. The 123 was prepared by mixing Y_2O_3, $BaCO_3$ and CuO, and annealed at 950°C for 8 h in a pre-heated oven, followed by oven cooling. A similar procedure was used for synthesizing 123F4, with BaF_2 replacing the $BaCO_3$. The 123F2 was prepared by mixing equimolar 123 and 123F4 and subjected to 8 h of annealing at 950°C, followed by oven cooling.

All weight loss curves were obtained using a Perkin-Elmer Thermogravimetric Analysis 7 (TGA-7) instrument. The TGA-7 is controlled by a TAC 7/7 controller and is coupled with a Perkin-Elmer 7700 Series data system. Heating and cooling rates of 5°C/min. were used, and the purging gases were of zero-grade from Air Products. The X-ray diffractograms were taken with a Rigaku XRD model CN4148H2 with 12kW rotating anode and computerized data system, including a JCPDS data base. The radiation source was CuK_{α} and the scanning rate (2θ) 5° per minute with an X-ray power of 150 mA at 50 kV.

RESULTS AND DISCUSSIONS

Fig. 1 shows the X-ray diffractograms of the $Y_1Ba_2Cu_3O_{7-x}$, $Y_1Ba_2Cu_3F_2O_y$, and $Y_1Ba_2Cu_3F_4O_y$. The bulk $Y_1Ba_2Cu_3O_{7-x}$ sample, noted as 123 for convenience, was found to be multiphasic, with the superconducting 123 phase as the major phase (Fig. 1). The impurities contained in this sample are Y_2BaCuO_5 and traces of $BaCuO_2$ and unreacted CuO. These impurities are common in this class of superconducting oxide. Based on the XRD data, the purity of the sample was estimated to be about 90%. From our previous study (19), the oxygen content of this sample was found to be greater than 6.9, suggesting an orthorhomic form of 123. The orthorhombic structure is reported to be the phase responsible for superconductivity above 90 K (2,20,21).

The XRD of $Y_1Ba_2Cu_3F_2O_y$ (123F2) indicates that the sample contained a high fraction of BaF_2, a significant amount of Y_2BaCuO_5, and unreacted CuO. The Y_2BaCuO_5 could be from the 123 sample, as the 123F2 was prepared by mixing the 123 and $Y_1Ba_2Cu_3F_4O_y$. A similar diffraction pattern was also observed on $Y_1Ba_2Cu_3F_xO_y$ (x = 1), prepared using fluorine gas treatment, by Cirrilio et al. (15) and Kim et al. (22) in their fluorinated sample. Even though the XRD is dominated by BaF_2 peaks, peaks that are consistent with the 123 superconducting phase were observed. The strongest orthorhombic 123 doublet peaks (2θ = 32.7° for 123 phase) were shifted to a lower 2θ value (32.6°), indicating an expansion of the crystal lattice. This lattice parameters expansion were also observed by several groups who reported successful incorporation of fluorine (12-14). The lattice expansion is believed to be the result of incorporation of fluorine into the crystal structure. It should be noted that a significant amount of BaF_2 was present in the sample, and therefore only a fraction of BaF_2 is reacted to form the fluorinated phase. The weight loss transition at around 480°C, 80°C higher than the 123 (Fig. 2), can be attributed to the more thermally stable fluorinated phase. From our XRD and previous TG result (19), we estimated that less than 20% of the fluorinated phase is present in the 123F2 sample.

The multiphasic $Y_1Ba_2Cu_3F_4O_y$ (123F4) is mainly made up of BaF_2, CuO, and Y_2O_3 (Fig. 1c). The amount of BaF_2 present is significantly greater than in the 123F2, as expected. As with 123F2, the XRD of the 123F4 sample also exhibit peaks that are consistent with the 123-phase. However, the doublet peaks of 123F2 and 123 at around $2\theta = 32.7^O$ become a single peak at a higher 2θ ($2\theta = 32.97^O$) in 123F4. This suggests that the phase is related to the amorphous-tetragonal fluorinated sample, which has smaller lattice parameters. A similar observation was reported by Lagraff et al. (23) for a sample that had been exposed for a long period of time to NF_3 gas. In addition, they observed a decreasing trend of the lattice parameters in their $Y_1Ba_2Cu_3O_{7-x}F_y$ as a function of nominal fluorine content. The 123F4 started to show a weight loss at around 920^OC (Fig. 2) in an oxygen-lean atmosphere. At this temperature the 123 sample is in a tetragonal form (19). This TG observation further supports the presence of amorphous-tetragonal structure.

CONCLUSIONS

Using XRD, we have determined that some of BaF_2, a precursor for $Y_1Ba_2Cu_3F_2O_y$ and $Y_1Ba_2Cu_3F_4O_y$ nominal composition, reacted to form fluorinated 123 phase in 123F2 and the amorphous-tetragonal fluorinated phase in 123F4. These two fluorinated phases are responsible for the difference in thermal behavior observed previously (19), which may account for the difference in superconductivity observed.

LITERATURE CITED

1. J.G. Berdnorz and Muller, A. K., Z. Phys. B, 64, 189 (1986).

2. Chu, C.W., Hor, P.H., Meng, R.L., Gao, L., Hung, Z.J., and Wang, Y.Z., Phys. Rev. Lett., **58**, 405 (1987).

3. Chen, J.T., Qian, L-X, Wang, L-Q, Wenger, L.E., and Logothetis, E.M., accepted Mod. Phys. Lett. B (1990).

4. Jostarndt, H.D., Galffy, M., Freimuth, A., and Wohlleben, D., Solid State Commun., **69**, 911 (1989).

5. Chen, J.T., Wenger, L.E., McEwan, C.J., and Logothetis, E.M., Phys. Rev. Lett., **58**, 1972 (1987).

6. Bourne, L.C., Cohen, M.L., Creager, W.N., Crommie, M.F., Stacy, A.M., and Zettl, A., Phys. Lett. A, **120**, 494 (1987).

7. Ovshinsky, S.R., Young, R.T, Allert, D.D., DeMaggio, G., and Vander Leeden, G.A., **Phys. Rev. Lett.**, **58**, 2579 (1987).

8. Bhargava, R.N., Herko, S.P, and Osborne, W.N., Phys. Rev. Lett., **59**, 1468 (19878).

9. Tyagi, A.K., Patwe, S.J., Rao, U.R.K., Iyer, R.M., Solid State Commun., **65**, 1149 (1988).

10. Hau, H.W., William, J.M., Kini, A.M., Kao, H-C I., Appelman, E.H., Chen, M.Y., Schlueter, J.A., Gates, B.D., Hallenbeck, S.L., Inorg. Chem., **27**, 5-8 (1988).

11. Meng, X.R., Ren, Y.R., Lin, M.Z., Tu, Q.Y., Lin, Z.J., Sang, L.H, Ding, W.Q., Fu, M.H., Meng, Q.Y., Li, C.J., Li, X.H., Qiu, G.L., and Chen, M.Y., Solid State Commun, **64**, 325 (1987).

12. Herrmann, R., Kubicki, N., Muller, H-U., Dwelk, H., Pruss, N., Krapf, A., and Rothkirch, L., Physica C, **153-155**, 936 (1988).

13. Nakayama, H., Fujita, H., Nogami, T., and Shirota, Y., Physica C, **153-155**, 936 (1988)

14. Fukushima, K., Kurayasu, H., Tanaka, T., and Watanabe, S., Jpn. J. Appl. Phys., **28**, L1533 (1989).

15. Cirillo, K.M., Wright, J.C., Seuntjens, J., Daeumling, M., and Larbalestier, D.C., Solid State Commun., **66**, 1237 (1988).

16. Sauer, N.N., Garcia, E., Martin, J.A., Ryan, R.R., Eller, P.G., Tesmer, J.R., and Maggiore, C.J., J. Mat. Res., **3**, 813 (1988).

17. Rao, U.R.K., Tyagi, A.K., Patwe, S.J., Iyer, R.M., and Sastry, M.D. Solid State Commun., **67**, 385 (1988).

18. Wang, H.H., Kini, A.M., Kao, H.I., Appleman, E.H., Thompson, A.R., Botto, R.E., Carlson, K.D., William, J.M., Chen, M.Y., Schlueter, J.A., Gates, B.D., Hallenbeck, S.L., and Despotes, A.M., Inorg. Chem., **27**, 5 (1988).

19. Kao, S., Lee, S., Ng, K.Y.S, Solid State Commun., **72**, 469 (1991).

20. Halasz, I., Jen, H.W., Brenner, A., Shelef, M., Kao, S., and Ng, K.Y.S., J. Solid State Chem., **92**, 327 (1991).

21. Halasz, I., Kischner, I., Projasz, T., Kovacs, Gy., Karman, T., Zsolt., G., Sukosd., Cs., Rozlasnik., N.N., and Kurti., J., Physica C, **153-155**, 379 (1988).

22. Kim, J.S., Swinnea, J.S., Manthiram, A., Steinfink, H., J. Solid State Commun., **66**, 287 (1988).

23. Lagraff, J.R., Behrman, E.C., Taylor, J.A.T., Rotella, F.J., Jorgensen, J.D., Wang, L.Q., and Mattocks, P.G. , Phy. Rev. B, **39**, 347 (1989).

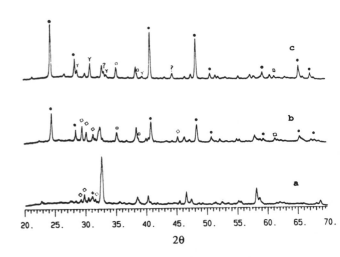

Figure 1. X-ray powder diffraction of $Y_1Ba_2Cu_3O_{7-x}$ (a), $Y_1Ba_2Cu_3F_2O_y$ (b), and $Y_1Ba_2Cu_3F_4O_y$ (c). Circles are diffraction lines for BaF_2, square for CuO, Y for Y_2O_3, cross for $BaCuO_2$, and diamond for Y_2BaCuO_5. Unmarked lines are from the 123 phase or fluorinated 123 phases.

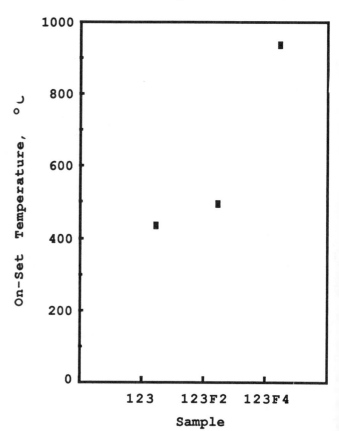

Figure 2. Weight loss transition temperature of $Y_1Ba_2Cu_3O_{7-x}$ (123), $Y_1Ba_2Cu_3F_2O_y$ (123F2), and $Y_1Ba_2Cu_3F_4O_y$ (123F4).

HIGH TEMPERATURE SUPERCONDUCTING THIN FILMS

Julia M. Phillips ■ AT&T Bell Laboratories, Murray Hill, NJ 07974

In the five years since the first report of successful high temperature superconducting film growth, considerable progress has been made. The plethora of techniques which have been explored have now been narrowed to a relatively small number which can produce films with state-of-the-art properties. While serious scientific and technological issues remain, the progress to date gives confidence that the remaining problems can be solved. One of the most important materials issues in film growth concerns the choice of substrate. The search for viable substrate materials is an active area of research. Significant breakthroughs are required to identify materials which satisfy all of the requirements which the thin films pose.

Thin films of high temperature superconducting (HTS) materials were attempted very soon after the materials were discovered [1]. Nearly every growth method which had been used for the fabrication of other types of thin films was tried by one group or another [2]. Most of the early results were very disappointing, with poor superconducting properties (except, possibly, for T_c), crystallinity, and morphology. In the five years since the first report of successful HTS film growth, considerable progress has been made. The plethora of techniques which have been explored have now been narrowed to a relatively small number which can produce films with state-of-the-art properties. While serious scientific and technological issues remain to be addressed, the progress to date gives confidence that the remaining problems can be solved.

One of the most important materials issues in HTS film growth concerns the choice of substrate. The best films which can currently be produced are grown on materials which have serious limitations. The search for viable substrate materials is an active area of research. It is fair to say that significant breakthroughs are required to identify materials which satisfy all of the requirements which the HTS films pose.

This paper begins by discussing the current status of HTS thin film growth, looking critically at the growth techniques which are presently receiving the most attention. Because the body of work which has been done on $YBa_2Cu_3O_{7-\delta}$ (YBCO) thin films is far larger than the work on all other HTS films combined, references in this discussion concern YBCO unless otherwise specified. In general, progress on YBCO has been much greater than on the growth of other HTS films. Because the issue of substrate materials is so critical for the ultimate optimization of HTS film quality, the second part of the paper focuses on substrate materials which have received the most attention. From this discussion, the need for further work on substrate materials becomes apparent.

HIGH TEMPERATURE SUPERCONDUCTOR THIN FILM GROWTH

General Methods of Film Growth

When considering the growth of HTS thin films, it is important to recognize that there are two major categories of growth techniques which can be utilized: *in situ* and *ex situ* film growth. *In situ* HTS films superconduct when they are removed from the growth chamber. *Ex situ* films do not superconduct after growth; they must be annealed in an oxygen-rich environment to become superconducting. This is frequently called a "post anneal". *In situ* and *ex situ* films have different growth modes. *In situ* films grow in a more or less layer-by-layer manner. As such, surface diffusion is particularly important, as the deposited atoms and molecules migrate to their equilibrium lattice sites. *Ex situ* films, on the other hand, are deposited in an amorphous state. Thus bulk diffusion leads to solid phase epitaxy which gives rise to the HTS crystal structure.

In situ films can generally be grown at lower growth temperature, due to the dominance of surface diffusion for achieving epitaxy. Multilayers are readily grown by such techniques. The best films have smooth, almost featureless, surface morphology. They also have high J_c, with limited degradation in a magnetic field, and low surface resistance. Precise control of the film stoichiometry is not generally required in order to obtain high quality films [3]; control of the cation composition to within 10% is usually sufficient.

While it is possible to grow *in situ* HTS films of extremely high quality, these growth techniques do have a number of disadvantages. The growth apparatus is generally very complex, given the requirements of substrate heating during growth and the presence of high partial pressures of oxygen. Films grown by some *in situ* growth techniques may have somewhat depressed T_c's, even though their other properties are excellent. In general, it is difficult to grow uniform, high quality, large area films using *in situ* techniques.

Ex situ films require a simpler growth apparatus than *in situ* films. Low oxygen pressure during film growth is generally acceptable. The best *ex situ* films have higher T_c than do some of the *in situ* films. Double sided growth, which is important for some applications which require a ground plane in addition to a device film, is easily accomplished, and the growth of large area uniform films is rather straightforward. The non-superconducting precursor films may, however, be unstable. Compared to *in situ* processes, higher processing temperature is required to render the films superconducting, since bulk diffusion processes must be activated. Multilayers are difficult or perhaps impossible to grow because of the necessity of an anneal after growth. The films generally do not have morphologies which are as good as the best *in situ* films. The best quality films are less than 200 nm thick, and even these have somewhat higher surface resistance than the best *in situ* films. Finally, the best *ex situ* films are stoichiometric to within 1% [4].

HTS film growth techniques can also be divided between those which use multiple sources and those which require only a single source. Multiple source evaporation offers the advantage of flexibility. Small changes in film composition are readily achieved in such a system. With good deposition monitoring and control, accurate control of composition is also possible. Atomic layering is also possible in principle, which allows the fabrication of artificially layered compounds which may not be stable in bulk form. Some people also believe that the control which is theoretically possible using multiple source growth is necessary to achieve the highest quality films. Needless to say, multiple source deposition systems tend to be extremely complex, especially if the issues of deposition monitoring and control are addressed properly. Accurate flux monitoring of each source is required. In addition, each source must operate stably so that the relative deposition rate of each material remains constant. Many parameters must be controlled. This frequently leads to a problem of reproducibility in multiple source systems.

Single source growth is much simpler, both in theory and in practice. The apparatus can obviously be much less complex and more inexpensive. Accurate monitoring of the deposition source is less critical, and source stability is less crucial than in multiple source growth. The existence of fewer critical parameters makes single source apparatus more reproducible, in general. On the other hand, fine tuning film composition is difficult, generally involving either the use of a new source material or the tuning of other deposition parameters which affect the film composition in complicated ways.

HTS thin films have been grown by both physical and chemical deposition techniques. Physical deposition involves the production of a vapor which includes only the species to be deposited (some combination of atoms, molecules and radicals), possibly mixed with an inert ambient (e.g. Ar). Chemical deposition can take either of two general forms. Chemical vapor deposition (CVD) involves passing a vapor of molecules containing the elements to be deposited (in addition to other elements, notably hydrogen and carbon). The conditions in the reactor lead to cracking of the molecules and deposition of the desired species. Condensed state chemical deposition involves the application onto a substrate of a precursor layer containing the elements of the film in the desired ratio, followed by an annealing sequence which drives off the undesired elements and leaves only the film constituents.

Physical deposition offers the advantage of having no extraneous elements present during film growth. Such additional species may lead to impurity incorporation in the film, which inevitably has a bad effect on film properties. In addition, a variety of physical deposition techniques have been demonstrated to give high quality HTS films. Without exception, however, physical deposition takes place in a vacuum system, which necessarily adds complexity to the deposition apparatus. Many physical deposition techniques are incompatible with high pressures of oxygen. This must be circumvented in order to obtain HTS films.

CVD is, on the other hand, compatible with hostile gases, including oxygen. In addition, it is a familiar technique in the thin film industry. CVD of HTS films, though, must make use of precursor molecules which have heavy elements in them (such as barium). This makes it difficult to obtain volatile precursors. In addition, the precursors must crack correctly and reproducibly in order to avoid film contamination. The entire issue of precursor materials has not yet been adequately addressed, so that the film grower must pay great attention to this aspect of CVD.

Condensed state chemical deposition is the simplest of the deposition techniques described here. Because macroscopic quantities of precursor species can be mixed, it is quite straightforward to achieve very accurate stoichiometric control of the precursor film, something which is much more difficult in vapor deposition. Because the precursors contain elements which must not be present in the final HTS film, selection of the correct chemistry is extremely important. Contamination is an issue in nearly all films grown by this general technique. Possibly for this reason, state-of-the-art films have not been grown by chemical deposition.

Specific Methods of HTS Thin Film Growth

One popular method of thin film growth which has been applied to HTS materials is evaporation [5-9]. In this technique, a separate source is required for each metal to be deposited. Because this is inherently a multiple source film growth technique, each source must have real time rate control of the deposition. The deposition rate is typically 0.1-1 nm/s.

Evaporation can be used in either the *in situ* or *ex situ* mode. If *ex situ* films are to be produced, a modest background pressure of 10^{-5}-10^{-6} Torr

oxygen is required. The most successful *ex situ* deposition technique is the "BaF_2 process", which uses BaF_2 as the Ba metal source in $YBa_2Cu_3O_{7-\delta}$ (YBCO) [5,6]. These precursor films are very stable in air, whereas precursor films which contain Ba metal tend to react to form carbonates which are difficult to dissociate during the post anneal. *Ex situ* films are generally deposited onto an unheated substrated.

The BaF_2 process requires a two-stage anneal after growth to render the HTS films superconducting. The first stage of the anneal is performed in a wet oxygen ambient in order to hydrolyze the BaF_2. This annealing stage, which is at elevated temperature ($\sim 850^{\circ}C$) is also when solid phase epitaxy occurs to give the tetragonal 123 YBCO phase. Subsequently, a lower temperature annealing stage in dry oxygen allows the full oxygenation of YBCO.

In situ films are deposited onto a heated substrate [7-9]. The substrate temperature is usually about $700^{\circ}C$ for c-axis oriented YBCO. The most difficult requirement for the deposition of *in situ* films is the necessary presence of oxygen during film growth. If molecular oxygen is used, the pressure must be in the 0.1-1 mTorr range in the vicinity of the substrate [7]. Such pressures are incompatible with evaporation sources, so that differential pumping must be used to protect them. Substrate heating is also difficult in so much oxygen. Various approaches to this problem include the use of Pt as a heating element, tungsten-halogen heating lamps, and the use of sealed heaters.

In order to avoid the necessity of differential pumping, some investigators use a reactive oxygen-containing molecule in place of O_2.

The most common examples are ozone and NO_2. In the case of ozone, the use of pure O_3 allows the pressure in the vicinity of the substrate to be lowered by approximately one order of magnitude over that allowed by O_2. Illumination of the substrate by ultraviolet light allows the pressure to be reduced by one more order of magnitude [8].

Other investigators have used activated oxygen to lower the pressure requirement [9]. For example, a remote microwave oxygen plasma source has been developed which supplies about 8% activated oxygen to the substrate region. This allows the substrate region to operate in the 10^{-3} Torr range, while the pressure in the main part of the chamber remains in the 10^{-5} Torr regime.

In its most sophisticated state, evaporation becomes molecular beam epitaxy [10]. Sources can be shuttered sequentially to enhance layer-by-layer growth, and artificially layered films of compositions not achievable in bulk form can be fabricated.

Another method of HTS film growth which has been used with considerable success is sputtering [11-13]. Sputtering can be performed using either multiple metal (or oxide) targets [11] or a single oxide target [12]. The general principles are the same in either case, but far better results have been obtained using a single target, so this discussion will focus on that variant of the technique. Basically, sputtering involves the use of energetic ions (such as Ar^+) to erode a target. The target species then deposit onto a heated substrate which is located near the target.

In on-axis sputtering, the substrate faces the target [13]. This gives reasonable deposition rates (≥ 0.1 nm/s), but this geometry places the target within the plasma region, so that the deposited film is bombarded by ions. This may result in film damage and, more importantly in the case of HTS films, preferential resputtering of some atomic species. The films deposited also tend to be quite nonuniform.

Because of these problems, off-axis sputtering has become quite popular for HTS film deposition [12]. In this configuration, the substrate is oriented at $90^{o}C$ with respect to the target, so that it is not in the plasma. This results in uniform and stoichiometric films with state-of-the-art properties. The major drawback of the off-axis geometry is the relatively low deposition rate, ~100 nm/hr.

Pulsed laser deposition (PLD, also known as laser ablation) has also been used to grow films of high quality [14]. In this technique, an excimer laser, focused to an energy density of ≥ 1 J/cm^2, impinges on a single target of the material to be deposited. The heated substrate faces the target, and ablated species, including atoms, molecules and radicals, are transferred to the substrate. For HTS film deposition, deposition typically takes place in ~100 mTorr of molecular oxygen. Laser ablation has been touted as allowing the stoichiometric transfer of atomic species from the target to the substrate. In reality, the stoichiometry of the deposited film depends on a number of parameters besides the target composition. These parameters include background gas pressure, laser energy density, and substrate temperature. Reasonable deposition rates of 0.1-0.5 nm/s are possible with laser ablation, but the area of uniform deposition is typically quite small. Another problem which is unique to this deposition technique is the frequent presence of particulates on the film. This is believed to arise from the use of a target which has non-uniform density, so that chunks of it may be blown off during the ablation process. The problem has largely been alleviated by the use of uniform high density targets.

CVD and metal-organic chemical vapor deposition (MOCVD) have been receiving increased attention for the growth of HTS films for the reasons mentioned above, in addition to the fact that it is quite straightforward to deposit large area films by this technique [15]. Metal-containing precursor molecules are volatilized and passed through the reactor in the correct proportion in an inert carrying gas. Oxygen is introduced separately. As mentioned above, the issues concerning precursors and possible contamination from them are just now being addressed. While films with high transition temperature and high critical currents have been produced, they tend to have worse morphology than state-of-the-art films and to have worse high frequency characteristics such as surface resistance.

Of the condensed state chemical deposition techniques which have been attempted, by far the most successful is the metallorganic deposition (MOD) method [16]. In this technique, a solution of metal trifluoroacetates is spin-coated onto the substrate. During an anneal at modest temperature, the precursor film is decomposed to form a metal oxyfluoride film. This is then heat treated in moist O_2 to remove the fluorine and enable the epitaxial growth of YBCO. Finally, the film is oxygenated in pure O_2. After

decomposition of the precursor film, the procedure is much like the BaF_2 process described above. The results are also similar. This is the only condensed state chemical deposition technique which has yielded films that are nearly as good as state-of-the-art films. Because of the inherent simplicity of this approach, there is considerable incentive to determine the limits to the film quality which can be produced.

Major Unresolved Issues

While tremendous progress has been made in HTS film growth over the past five years, there are still a number of unresolved issues which must be addressed. In the case of *ex situ* film growth, the BaF_2 process is the only physical deposition process which is still being pursued seriously. While films grown by this technique have excellent superconducting properties generally, the best films tend to have weak flux pinning, so that their current carrying capability in a magnetic field is poor. For some device applications, this can actually be an advantage, but for many it is a serious problem. Recent experiments involving performing the post anneal in reduced oxygen partial pressure have suggested that this limitation is not inherent in the *ex situ* growth process. A more serious problem with *ex situ* films is that multilayer growth has not been demonstrated, except for growth of a single film on each side of a substrate. Unless this difficulty is overcome, this may limit *ex situ* growth to certain applications niches.

In situ evaporation has two major issues. The first of these is the necessity of supplying large amounts of oxygen during growth. A second, perhaps more serious problem, is that of growing large area uniform films.

The current goal is to grow high quality films on 4" diameter substrates. This will require significant engineering advances to accomplish.

On-axis sputtering has two major problems. First, as mentioned above, the composition of the HTS films is frequently not very stoichiometric or uniform due to the sputtering geometry. This has led to the second problem, namely reduced film quality compared with what can be produced by other techniques.

Off-axis sputtering is capable of growing extremely high quality, smooth films. Its only major disadvantage at present is the low growth rate. This has been alleviated to some extent by going to a cylindrical sputtering geometry, but this approach has its own limitations. The targets for this geometry are very expensive, and the sputtering parameters do not scale exactly from off-axis sputtering, contrary to predictions.

PLD is also capable of growing very good films. The issue of particulates in the films has been of concern, but it seems to be soluble by using targets of sufficiently high quality. Growth of large area, uniform films is an issue which has not yet been solved by this growth technique.

Films grown by MOCVD have excellent superconducting properties, except for rougher film morphology and larger surface resistance than the best films grown by other techniques. Because of the advantages of the technique, such as the capability of large area growth and its familiarity in industry, these problems are receiving considerable attention.

Similarly, MOD is very attractive because of its simplicity. Major issues which must be resolved are the production of films of quality equal to those achievable by other techniques. Most notably, the film morphology must be improved.

SUBSTRATES FOR HIGH TEMPERATURE SUPERCONDUCTING THIN FILMS

Issues in HTS Thin Film Substrates

The quest for substrate materials that are capable of supporting excellent films of HTS materials has been in progress for nearly as long as HTS thin films have been prepared [17]. The list of desirable substrate properties contains a number of entries that are common to good substrates for essentially all classes of thin films. A good thermal expansion match is necessary, whether or not one is dealing with an epitaxial system. In the case of YBCO, this requirement is particularly important because of the brittleness of the superconductor. Thus, YBCO films on bulk Si substrates, even with a buffer layer of, say yttria stabilized zirconia (YSZ), develop cracks if they are over 50 nm thick, due to the large difference between the coefficients of thermal expansion of Si and BYCO [18]. Films on YSZ/Si layers on bulk sapphire do not have the same problem.

The best HTS films grown to date, as determined by a multitude of metrics including critical current density, morphology, and stability over time, are epitaxial on their substrates. This most likely dictates that the lattice mismatch between the film and substrate should be as small as possible, although high quality films have been grown on MgO, which has a mismatch of >9% [19].

Of particular concern for the growth of high quality YBCO films, whether or not they are epitaxial, is the chemical compatibility of the film with the substrate material. The constituents of YBCO are reactive with many substrates that might otherwise be good candidates (such as unbuffered Si) [20]. The relatively high temperatures required for growing even *in situ* films ($\geq 700^{\circ}$C) [2] make the compatibility requirement more severe than it would be if high quality films could be grown at lower temperature. In the case of *ex situ* films, the problem is even more severe, since the maximum temperature that the film/substrate couple must withstand is usually $\sim 850^{\circ}$C [5]. The issue of chemical compatibility has generally meant that the substrates that support reasonably high quality YBCO films are themselves oxides.

An ideal substrate would have a flat surface and be free of twins and other structural inhomogeneities, although a number of materials in current use as YBCO substrates do have such problems [21]. It would be desirable, at the very least, to grow films on a substrate that has no phase transitions within the temperature regime required for film processing. In the case of microwave applications, where the dielectric properties of the substrate have an important effect on device performance, the existence of a twinning transition in the processing range is entirely unacceptable, since it precludes device modeling [22].

Device applications impose a number of other property requirements on the HTS substrate material. Microwave applications are not generally very sensitive to the dielectric constant of the substrate (as long as it is uniform and, preferably, isotropic), but they do depend on having a low value of the loss tangent. This precludes the use of materials which contain magnetic ions, such as most of the rare earths. On the other hand, high frequency applications, e.g. interconnects, require a low dielectric constant [23]. The search for substrates that can support the growth of high quality epitaxial YBCO films has centered on materials having the perovskite crystal structure, usually oxides [1, 24-26].

$SrTiO_3$ saw early success as a substrate material [1], which is not surprising in view of its rather small lattice mismatch with YBCO and its ready availability. The prohibitively large dielectric constant of this material ($\varepsilon = 277$ at room temperature), coupled with its unavailability in reasonable sizes, has spurred the search for alternatives.

MgO has received a good deal of interest in light of its ready availability and its modest dielectric constant ($\varepsilon = 9.65$) [19]. As mentioned above, it has a 9% lattice mismatch with YBCO, but with proper substrate preparation, high quality epitaxial films can be grown. Even the best films, however, tend to have some high angle grain boundaries, which is probably responsible for the fact that such films have poorer high frequency characteristics than do the best films grown on other substrates. MgO also reacts with water vapor, which precludes its use as a substrate for films grown by the BaF_2 process or by MOD.

Sapphire (Al_2O_3) is of considerable interest as a substrate in view of its modest dielectric constant ($\varepsilon = 9.34$) and its commercial availability in large diameter wafers. Sapphire reacts with YBCO, however, so that a buffer layer (such as MgO or $LaAlO_3$) must be used in order to obtain high quality films [7]. Because the crystal structure is not cubic, the dielectric properties of sapphire are anisotropic, which makes the modeling of microwave device performance difficult.

In spite of serious problems with chemical reactivity, Si has received considerable attention as a substrate material due to the tantalizing possibilities of integrating semiconductors with superconductors. The best results which have been achieved to date have involved the use of YSZ as a buffer layer [18]. Even when this is done, the thermal mismatch of Si and YBCO is a serious problem which limits the ultimate YBCO thickness which can be grown.

$LaGaO_3$ was identified as a potential substrate material rather early [24]. Its lattice matches and thermal expansion match with YBCO are quite good, and its dielectric constant at room temperature is smaller than that of $SrTiO_3$ by one order of magnitude ($\varepsilon = 25$). There were early reports of the growth of high quality films on this substrate. However, $LaGaO_3$ has one serious drawback, namely its first order phase transition at 420K [21], well within the processing region for HTS films grown by any technique. This transition gives rise to steps on the surface of the substrate, which can be particularly detrimental in the case of very thin or patterned films.

$LaAlO_3$ ($\varepsilon = 23$) also offers small lattice and thermal mismatches with YBCO [25]. It, too, has a phase transition within the film processing regime (at 800K), but this transition is second order [21]. While this leads to substrate twinning, there is no discontinuous volume change, and surface steps are not a problem. The twinning has not prevented the growth of high quality films on this substrate [25], but it does make the fabrication of complex microwave devices such as filters impossible, since the dielectric properties of the substrate vary from point to point in a manner that cannot be controlled or predicted.

$NdGaO_3$ ($\varepsilon = 20$) was introduced as a possible alternative to the La-containing perovskites [26]. It has smaller lattice mismatches than either $LaGaO_3$ or $LaAlO_3$. It also has the advantage of having no phase transitions between its melting point and room temperature, so twin-free substrates are available for YBCO growth. Nd^{3+} is a magnetic ion, however, which precludes its use as a substrate for microwave devices. There have been conflicting reports regarding the quality of film which can be grown on this substrate.

SUMMARY

There has been tremendous progress in the growth of HTS thin films over the five years since their first fabrication. A few techniques have emerged as being particularly promising for the growth of films with optimized properties. Each of these techniques has its own set of advantages and disadvantages, so that it is impossible to predict which one(s) will ultimately become the most popular. A speedy resolution of this issue will depend on the existence of a technological pull by applications which require a defined set of film properties.

Progress in the identification and utilization of substrate materials for HTS films has also been substantial, but less impressive than progress in the films themselves. Considerable work is needed, since some of the remaining problems in the films are almost certainly connected with the lack of an ideal substrate on which to grow them. Work continues in this field, and it is likely that much better substrate materials will be available in a year or so than can be obtained today.

REFERENCES

1. Chaudhari, P., R. H. Koch, R. B. Laibowitz, T. R. McGuire, and R. J. Gambino, Phys. Rev. Lett. 58, 2684 (1987).

2. Humphreys, R. G., J. S. Satchell, N. G. Chew, J. A. Edwards, S. W. Goodyear, S. E. Blenkinsop, O. D. Dosser, and A. G. Cullis, Supercond. Sci. Technol. 3, 38 (1990).

3. Kwo, J. R., private communication.

4. Carlson, D. J., M. P. Siegal, J. M. Phillips, T. H. Tiefel, and J. H. Marshall, J. Materi. Res. 5, 2797 (1990).

5. Mankiewich, P. M., J. H. Schofield, W. J. Skocpol, R. E. Howard, A. H. Dayem, and E. Good, Appl. Phys. Lett. 51, 1753 (1987).

6. Siegal, M. P., J. M. Phillips, Y.-F. Hsieh, and J. H. Marshall, Physica C 172, 282 (1990).

7. Sliver. R. M., A. B. Berezin, M. Wendman, and A. L. de Lozanne, Appl. Phys. Lett. 52, 2174 (1988); Berezin, A. B., C. W. Yuan, and A. L. de Lozanne, Appl. Phys. Lett. 57, 90 (1990).

8. Siegrist, T., D. A. Mixon, and E. Coleman, submitted to Appl. Phys. Lett.

9. Kwo. J. R., M. Hong, D. J. Trevor, R. M. Fleming, A. E. white, R. C. Farrow, A. R. Kortan, and K. T. Short, Appl. Phys. Lett. 53, 2683 (1988).

10. Eckstein, J. N., I. Bozovic, K. E. von Dessonneck, D. G. Schlom, J. S. Harris, Jr., and S. M. Baumann, Appl. Phys. Lett. 57, 933 (1990).

11. Allen, L. H., E. J. Cukauskas, P. R. Brousard, and P. K. Van Damme, IEEE Trans. Magnetics 27, 1406 (1991).

12. Eom, C. B., J. Z. Sun, B. M. Lairson, S. K. Streiffer, A. F. Marshall, K. Yamamoto, S. M. Anlage, J. C. Bravman, T. H. Geballe, S. S. Laderman, R. C. Taber, and R. D. Jacowitz, Physica C 171 354 (1990).

13. Kawasaki, M., S. Nagata, Y. Sato, M. Funabashi, T. Hasegawa, K. Kishio, K. Kitazawa, K. Fueki, and H. Koinuma, Jpn. J. Appl. Phys. 26, L736 (1987); Li, Q., X. X. Xi, X. D. Wu, A. Inam, S. Vadlamannati, W. L. McLean, T. Venkatesan, R. Ramesh, D. M. Hwang, J. A. Martinez, and L. Nazar, Phys. Rev. Lett. 64, 3086 (1990); Adachi, H., K. Hirochi, K. Setsune, M. Kitabatake, and K. Wasa, Appl. Phys. Lett. 51, 2263 (1989).

14. A. Inam, M. S. Hegde, X. D. Wu, T. Venaktesan, P. England, P. F. Miceli, E. W. Chase, C. C. Chang, J. M. Tarascon, and J. B. Wachtman, Appl. Phys. Lett. 53, 908 (1988).

15. Hiskes, R., S. A. DiCarolis, J. L. Young, S. S. Laderman, R. D. Jacowitz, and R. C. Taber, Appl. Phys. Lett. 59, 606 (1991).

16. McIntyre, P. C., M. J. Cima, J. A. Smith, Jr., R. B. Hallock, M. P. Siegal, and J. M. Phillips, J. Appl. Phys. 71, 1868 (1992).

17. See, for example. Gurvitch, M. and A. T. Fiory, Appl. Phys. Lett. 51,

1027 (1987).

18. Fork, D. K., F. A. Ponce, J. C. Tramontana, N. Newman, J. M. Phillips, and T. H. Geballe, Appl. Phys. Lett. *58*, 2432 (1991).

19. Li, Q., O. Meyer, X. X. Xi, J. Geerk, and G. Linker, Appl. Phys. Lett. *55*, 310 (1989); Xi, X. X., G. Linker, O. Meyer, E. Nold, B. Obst, F. Ratzel, R. Smithey, B. Strehau, F. Weschenfelder, and J. Geerk, Z. Phys. B *74*, 13 (1989); Moeckly, B. H., S. E. Russek, D. K. Lathrop, R. A. Buhrman, J. Li, and J. W. Mayer, Appl. Phys. Lett. *57*, 1687 (1990).

20. Mogro-Campero, A., B. D. Hunt, L. G. Turner, M. C. Burell, and W. E. Balz, Appl. Phys. Lett. *52*, 584 (1988); Madakson, P., J. J. Cuomo, D. S. Yee, R. A. Roy, and G. Scilla, J. Appl. Phys. *63*, 2046 (1988).

21. O'Bryan, H. M., P. K. Gallagher, G. W. Berkstresser, and C. D. Brandle, J. Mater. Res. *5*, 183 (1990).

22. Lyons, W. G., R. S. Withers, J. M. Hamm, A. C. Anderson, D. E. Oates, P. M. Mankiewich, M. L. O'Malley, R. R. Bonetti, A. E. Williams, and N. Newman, Proceedings of the Fifth Conference on Superconductivity and Applications, T. H. Kao, H. S. Kwok, and A. E. Kaloyeros, editors (American Institute of Physics, New York, 1992), vol. 251, p. 639.

23. G. Arivalingam, private communication.

24. Sandstrom, R. L., E. A. Giess, W. J. Gallagher, A. Segmuller, E. I. Cooper, M. F. Chisolm, A. Gupta, S. Shinde, and R. B. Laibowitz, Appl. Phys. Lett. *53*, 1874 (1988).

25. Simon, R. W., C. E. Platt, A. E. Lee, K. P. Daly, M. S. Wire, J. W. Luine, and M. Urbanik, Appl. Phys. Lett. *53*, 2677 (1988).

26. Koren, G., A. Gupta, E. A. Giess, A. Segmuller, and R. B. Laibowitz, Appl. Phys. Lett. *54*, 1054 (1989).

Index